Hannah's Heirs

Hannah's Heirs

The Quest for the Genetic Origins of Alzheimer's Disease

EXPANDED EDITION

DANIEL A. POLLEN, M.D.

New York Oxford
OXFORD UNIVERSITY PRESS

Oxford University Press

Oxford   New York
Athens   Auckland   Bangkok   Bogotá   Bombay
Buenos Aires   Calcutta   Cape Town   Dar es Salaam
Delhi   Florence   Hong Kong   Istanbul   Karachi
Kuala Lumpur   Madras   Madrid   Melbourne
Mexico City   Nairobi   Paris   Singapore
Taipei   Tokyo   Toronto

and associated companies in
Berlin   Ibadan

First published in 1993 by Oxford University Press, Inc.,
198 Madison Avenue, New York, NY 10016

First issued as an Oxford University Press paperback, 1996

Oxford is a registered trademark of Oxford University Press

Library of Congress Cataloging-in-Publication Data
Pollen, Daniel A.
Hannah's heirs : the quest for the genetic origins of
Alzheimer's disease /
Daniel A. Pollen.
p.   cm.,   Includes bibliographical references and index.
ISBN 0-19-506809-2
ISBN 0-19-510652-0 (Pbk.)
1. Alzheimer's disease—Genetic aspects—Case studies.
2. Alzheimer's disease—Popular works.
3. Alzheimer's disease—Pathogenesis.
I. Title   RC523.2.P65   1993
616.8'31042—dc20   93-66

1 2 3 4 5 6 7 8 9

Printed in the United States of America

*This book is dedicated to
the descendants of Hannah and Shlomo
even unto the sixth generation
and especially to their grandson Charles
and great-grandson Ben,
physicians both,
who pointed the way.*

Preface

This is a story about despair, tragedy, and death, but it is also about hope, triumph, and survival. Above all, it is the story of the remarkable courage of a few individuals from among the many at risk for familial Alzheimer's disease. In taking an active part in the making of science, they have become a part of the history I tell here—a triumph of human mettle against one of the cruelest and most capricious of adversities, one that seals the fate of an individual at the moment of conception.

This is an astonishing story, the more so because these events have already taken place and yet the story continues. The names of patients have been omitted or changed in the interest of privacy and medical confidentiality. My own involvement began quite by accident when a patient came to my clinic bearing a single sheet of paper documenting the occurrence of Alzheimer's disease in his family over generations (called a pedigree). The men and women represented by the squares (male) and circles (female) on this document, in the convention of the geneticist, were darkened for the presence of the disease or left open for those who had either escaped the disease or were not yet affected. At the time I realized only that somewhere along the 3 billion base-pairs of their DNA lay hidden the secret of a genetic defect that inflicts the awful suffering of familial Alzheimer's disease which strikes the bearers in their prime, in middle life. Could this defect also provide a clue to the molecular pathology of the more frequent late form of the disease which strikes the aging population?

That a member of this family had come to my neurology clinic was ironic. My own research was in the physiology of the visual cortex, an area remote from molecular genetics, and at the time I saw myself as only an attentive bystander who would make sure that the pedigree was put in the right hands of an experienced geneticist.

In time, molecular geneticists would link the first known genetic defect in familial Alzheimer's disease to a particular chromosome—the long arm of chromosome 21—and begin the arduous search to locate the precise abnormal gene causing the disease in some families. With this first step scientists would be that much closer to devising therapy for Alzheimer's. At this moment, molecular geneticists throughout the world are carrying out the research that all of us hope will help prevent this disease. Here I provide only the highlights of a story which is still unfolding. But as a neurologist, working with patients and their families, I discovered that there was a parallel story of compelling interest to tell. My colleagues in molecular genetics dealt with the DNA, the basic instruction set of heredity, that was sampled from my patients and their unaffected relatives. I had the privilege of working with these individuals who, by the accident of heredity, became the participants on whose lives this story is based. The family with whom I worked is part of a much larger story, a saga of families and scientists from Moscow in the East to Seattle in the West, from Kuopio in the North of Finland to Calabria in the South of Italy, all working toward a common goal.

In the course of telling this story, I have tried to cover some of the epic discoveries in the annals of genetics and neurology that have made possible the present quest but to do so at a level understandable to the general reader. I have tried to explain these advances as simply as possible without violating the integrity and accuracy of the subject matter. My basic premise is that Alzheimer's disease has affected or will affect the lives of many who are dear to each of us. From this premise stems my conviction that all of us, general readers and scientists alike, will have gained from looking back at how we have reached the present high ground and from looking forward to our eventual objectives. Surely, I will not have succeeded in explaining every discovery to the satisfaction of every reader. Where I may have failed, I ask the reader's forbearance and urge him or her to "read past" a technical point here and there and join with me in the drama of the larger story.

Finally, although I have dedicated this book to a remarkable family at risk for Alzheimer's disease, I hope that this book will inspire others to action who now live in the shadow of some of the over 4,000 or so known genetic diseases that afflict humankind. For although the diseases may differ, the strategies to be followed in seeking the cause and eventual treatment follow along similar paths.

Worcester, Massachusetts D.A.P.
October 1992

Acknowledgments

I am very grateful to my wife Linda, Jeff Resnick, and Dr. Ann Mitchell for reading drafts of my earliest chapters and providing me with helpful suggestions. Their enthusiasm for the project fueled my early efforts.

Throughout the course of my writing this book I have questioned and received assistance from many people either by personal interview, phone, or letter. Others have assisted in the research about which I write. These include Drs. Paul Brown (Bethesda), Robert Butler (New York), the late Mrs. Bobbie Glaze Custer (Minneapolis), Tom Daniels (Salt Lake City), Dennis Dickson (New York), David Drachman (Worcester), Louis DeGennaro (Worcester), Robert Feldman (Boston), Patrick Fox (San Francisco), Carlton Gajdusek (Bethesda), Pierluigi Gambetti (Cleveland), Svetlana Ivanovna Gavrilova (Moscow), Marc Godek (Bethesda), Peter Goodfellow (London, England), Jaap Goudsmit (Amsterdam), Tibor Greenwalt (Milwaukee), James Gusella (Boston), Aaron Gutman (Kaunas), William Hadlow (Hamilton, Montana), Jonathan Haines (Boston), James Hamos (Worcester), John Hardy (London, now at Tampa), Leonard Heston (Min-neapolis), David Housman (Cambridge, USA), Paul Hoff (Munich), Egidijus Jarzhemskas (Kaunas), Modest Kabanov (Leningrad, once again St. Petersburg), Robert Katzman (La Jolla), Hans Lauter (Munich), Carol Lippa (Worcester), Michael Litt (Portland, Oregon), Hans-Dieter Lux (Munich), Vitezslav Orel (Brno), Brian O'Donnell (Worcester), Tony Phelps (Chicago), Ronald Polinsky (Bethesda), Stanley Prusiner (San Francisco), Allen Roses (Durham), Alexander Rich (Cambridge, USA), Peter St. George-Hyslop (Boston, now at Toronto), Gerard Schellenberg (Seattle), Dennis Selkoe (Boston), Marrott Sinex (Boston), Ellen Solomon (London), Gerry Stone (Chicago), Joan Goodwin Swearer (Worcester), Robert Terry (La Jolla), Rudi Tanzi (Boston), Allen Tobin (Santa Monica), Joseph Tonkonogy (Kiev, Leningrad, now at Worcester), Christine Van Broeckhoven (Antwerp), Marat Vartanyan (Moscow), Nina Ivanovna Voskreskanskaya (Moscow), James Weber (Marshfield, Wisconsin), Nancy Wexler (New York), Lorna Wheelan—now Lady Hill (London), and Bruce Yankner (Boston).

I received constant encouragement, advice, and much help from many of Hannah's descendants whose names remain anony-mous only to protect the medical confidentiality of their afflicted or at-risk relatives. I have a very special debt of gratitude to Linda Curnin who typed the manuscript and worked through the many revisions with patience, dedication, support, and many helpful sug-gestions.

I gratefully acknowledge the numerous constructive criticisms by the editorial staff of Oxford University Press and their desig-nated outside reviewers. I especially thank Joan Bossert, Editor at Oxford University Press, who shared my conviction that this story could be told clearly and accurately for all readers without violating the scientific integrity of the subject matter. Her keen and skillful editorial assistance moved this goal from concept to reality. I espe-cially thank Victor McKusick (Baltimore) who reviewed the origi-nally submitted manuscript and with a wise nod sent it forward for further scrutiny to Robert Cook-Deegan (Washington) who meticu-lously reviewed both the original and penultimate drafts. His inci-sive criticisms and comments with respect to style, structure, and content have led to a vastly improved final version. I also sincerely thank Peter Whitehouse (Cleveland), who reviewed the manuscript specifically to assure the accuracy and readability of the sections on the neurology and molecular genetics of Alzheimer's disease, for his extremely important and helpful suggestions. I also express my sincerest thanks to Betty Pessagno who carefully edited and fine-tuned the final draft. Even so, despite the intense interest by others in ensuring the accuracy of this endeavor, the final responsibility for any unrecognized errors remains mine alone.

I also thank Dr. Whitehouse's colleague, Dr. Stephen Post (Cleveland), a biomedical ethicist at Case Western Reserve University's Alzheimer's Center, who along with both Peter Whitehouse and Robert Cook-Deegan encouraged me to expand certain chapters so as to address sensitive ethical issues that have arisen in the course of research projects in human molecular genetics, where in certain instances the best interests of patients and those of researchers have been in conflict.

Finally, I honor the memory of my late parents, David and Anne Pollen, who did not live to see the completion of this work but who, in their prime, had prepared me for the task.

Contents

PART III
THE ASCENT

PART IV
THE PATH AHEAD

EPILOGUE

Hannah's Heirs

From past generations we receive a few strands of DNA,
sometimes a heritage,
a memory of one sort or another.

Prologue

About noon on May 6, 1985 Dr. David Drachman, chairman of the Department of Neurology at the University of Massachusetts Medical School, asked if I would see a patient for him that afternoon in my clinic. Sue Boiteau, the scheduling secretary, had just informed him that he was double-booked at 1 P.M. since she had committed him to see an emergency consult. My anticipated free hour of time had vanished.

A short note and some records including a genetic history from a physician-relative accompanied a robust-looking gentleman of fifty-one and his troubled-looking wife. In the setting of a busy clinic, a doctor seldom has enough time to review all documents before seeing a patient. I could see the paperwork later. Jeff's wife explained that her husband's recent memory had been declining over the last three or four years, that he had become more irritable with his children, and that he just didn't seem as sharp as he had once been. Jeff's wife, Susan, feared that her husband had developed Alzheimer's disease, a slowly progressive degenerative disease of the brain that first impairs and eventually destroys memory, especially recent memory, and all higher thought processes.

From the patient's genetic tree (see facing page), it was clear that Jeff was genetically at risk for Alzheimer's disease. Such people, both those destined to develop the disease and those destined to escape it, often live with an almost incapacitating fear as they enter the age of risk. My desire certainly was not to try to diagnose Alzheimer's disease, but exactly the opposite. My responsibility was to determine whether the patient actually had a dementia or whether his mental function was compromised as a result of depression or anxiety. If he had a dementia, then my obligation was to search for all possible treatable causes. Only if no treatable causes were found would I have to consider other potential diagnoses that included Alzheimer's disease as the statistically most likely.

As I took the history, it became clear that Jeff, though fully independent in his activities of daily living, had a number of worrisome problems. He enjoyed his hobbies, especially golf, and played as good a game as ever, but he could no longer keep the score. He had spent the winter in Florida but had failed the written part of the Florida driving test twice before finally passing it. He had shared in the driving rotation during the auto trip north but generally became irritable during his second hour at the wheel and occasionally drove through red lights. His wife reported that he had shown excessive irritability at restaurants when he was presented with checks he felt had been miscalculated. His wife had managed the family finances for the past five years and doubted that Jeff could handle them now. Finally, when they reached their summer home in New England, Jeff was for the first time unable to reconnect his television set.

As my examination of Jeff continued, it became evident that there were indeed definite impairments in his recent memory and in every cognitive function that I evaluated. For example, Jeff remembered that he had driven from Florida to Massachusetts along route I-95, but he could not recall the names of several states along the route. He could not draw a simple outline of the United States, nor could he locate accurately the position of major cities. His ability to calculate and his language skills were mildly impaired. His ability to interpret proverbs and his judgment were also mildly impaired. Apart from the mental status examination, the remainder of the neurologic examination was entirely normal. There was no evidence of a stroke, brain tumor, or any other condition that might produce a similar picture. This patient had as dif-

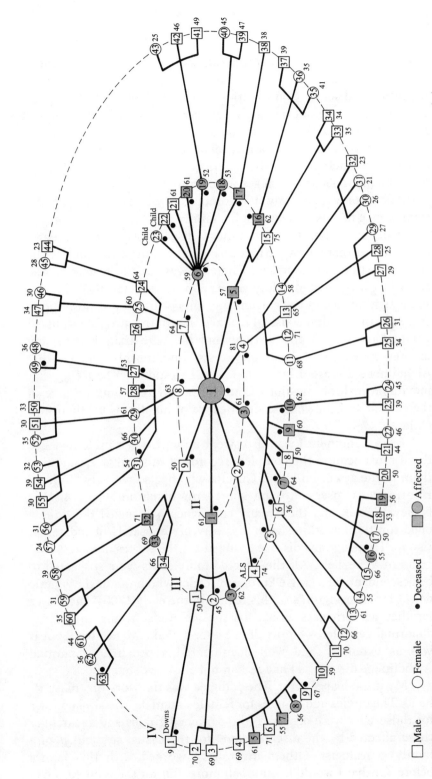

The original family tree that Jeff brought with him on the afternoon of May 6, 1985. This pedigree diagram represented the status of Hannah's family with respect to the incidence of familial Alzheimer's disease as of February 1981. Subsequently, the family tree would be enlarged to include 92 members of the fifth generation and a sizable number of youthful members of the sixth generation. The numbers just outside of each circle and square indicate the ages of the respective persons in 1981.

☐ Male   ◯ Female   ● Deceased   ⬤ Affected

fuse a decline in intellectual function as I had ever seen. His mental status examination could not have been more typical of Alzheimer's disease. Confirmation of the diagnosis would require blood work to rule out metabolic, endocrine, and infectious causes, and a CAT scan of the head to rule out hydrocephalus or lesions.

These would be done, but the chances of finding an alternative diagnosis in such a physically healthy, relatively young man apart from the mental status changes were remote.

An instant later, my jaw dropped open as I realized the significance of this patient and his family tree for future studies of the genetics of familial Alzheimer's disease. At this moment I remembered that piece of paper I had hastily put in the back of the chart less than an hour ago. There it was. In the format of the geneticist was a record of four generations of Jeff's family. The awesome significance of the existence of this pedigree was now apparent. Almost simultaneously, I realized that in the next 30 minutes I would have to tell this patient and his wife that, barring any unforeseen result of laboratory tests, he probably had familial Alzheimer's disease. I knew, too, that I would soon be facing that inevitable rush of questions from the family regarding the fate of the patient and the genetic significance of his illness for their four children.

At that moment I was also swept up with a certain knowledge that I must acquire the complete records of this pedigree and bring them to the attention of Jim Gusella who had so recently "linked" Huntington's disease to the short arm of chromosome 4. Jim, Nancy Wexler, and their many colleagues had seized the unique opportunity of working with a family of thousands of Indians at risk for Huntington's disease living along the shores of Lake Maracaibo in Venezuela. All the affected individuals were suspected to be descendants of a single European settler who brought the genetic defect for Huntington's to this region in the early 1700s. Such large families are priceless assets for the geneticist trying to find an abnormal gene. Jeff's family didn't offer a "Lake Maracaibo," but it was as extensive and well documented a pedigree of familial Alzheimer's disease as I had reason to believe existed.

My mind leaped on it. The pedigree was the work of a master's hand. The circles and squares for female and male, the slashed lines for those who were deceased, the darkened circles and squares for those affected by the disease, were not the casual scribblings of a family genealogist. Others must have worked with this family before. Of that I would learn much more. But for the moment I had to hold these thoughts back and address the concerns of the patient and his wife. Jeff's pedigree and his diagnosis offered momentous opportunity for research, but for Jeff and Susan the diagnosis confirmed the enormous personal tragedy they had long feared.

The family then asked for my provisional diagnosis. I gave it, as gently as I could. An awful silence filled the room. Susan spoke first, "Doctor, I guess we've known this for some time." She then spoke of Jeff's mother Sarah who had died with Alzheimer's at age 53. Susan had known Sarah in her final years and knew what she and Jeff would now have to face. Twenty-five years ago Susan and Jeff had often discussed whether they ought to have children. But then they were both young. Surely medicine would find a cure by the time Jeff reached the age of risk. And, perhaps Jeff would escape the disease. But this had not happened. And now Jeff and Susan spoke of their four children each of whom, it was now virtually certain, would carry a 50 percent chance of developing the disease 20 to 30 years hence. That is, unless something could be done.

I then explained to the family the unique opportunity for them to help in the search for the abnormal gene. Could the results come in time to help Jeff? Probably not, but perhaps they could come in time to help their children. Long before the dreams of a genetic therapy for a disease can be realized and even before a gene is completely located and identified, there comes a period when carrier identification and prenatal diagnosis become possible with ever-increasing accuracy. Such identification of presymptomatic carriers of Huntington's disease was already becoming available. Jeff spoke next, "Doctor, we will help in any way we can. We only want to help our children." So too, as I would later learn, would Jeff's more than 100 relatives known to be living in North America.

Thus, this day which had begun with a scheduling error had ended with a momentous event—information that would help molecular geneticists to find an abnormal gene that causes familial Alzheimer's disease was literally placed in my hands. Yet this day also was a historical crossroad. It is no exaggeration to say that the efforts of tens of thousands of lives had converged to make possible the research that could now begin on the molecular origins of Jeff's illness. I could not then have even dreamed of the rich legacies from the distant, as well as the recent, past that had made the day of Jeff's visit not one of despair alone but of hope as well. But a crossroad signifies a future as well as a past. The events that have taken place since Jeff's visit have been both extraordinary and unanticipated. No one could have imagined the diverse ways by which Jeff's visit has touched so many lives. And the story is still unfolding.

PART I
Founders

The Life and Death
of Hannah

"Very deep is the well of the past." So Thomas Mann begins his
quartet "Joseph and His Brothers." The present story also had its
beginnings very long ago, but thus far I can trace it back with
certainty only to the birth of Hannah in 1844.[1] Hannah's husband, Shlomo, was born in Bobruisk, in the Mogilev Oblast,
Byelorussia, or "White Russia." Some of her descendants assume
that she also was born in Bobruisk, but no one in America now
living knows for sure. A surviving granddaughter vaguely recol-
lects hearing as a child that Hannah may have come to Bobruisk
from Riga, Latvia, some 325 miles to the northwest.

Hannah spoke the Lithuanian dialect of Yiddish. Whether
born in Riga or Bobruisk, those of her ancestors who were Jewish
had probably come to these lands from Germany with the east-
ward migration of Jews, many as international traders, beginning
toward the end of the ninth century.[2] Perhaps later, Jews from
the even earlier settlements along the Black Sea, the Crimea and
Khazar kingdoms, migrated up into Eastern Europe. But there
can be no certainty that all of Hannah's ancestors were Jew-

ish. The ancient Russian state (Kievan Rus) was also forming during the ninth century and eventually unified the Eastern Slavic tribes out of which the Byelorussian, Ukrainian, and Russian states would one day be formed.[3] During the second half of the thirteenth century, as the Slavic lands faced both the Mongol Tatar invasions from the east and pillaging raids from the Teutonic Knights from the West, the princes of Lithuania seized the Western Rus. By the early fourteenth century Lithuania ruled from the Baltic to the Black Sea. By the end of the century, Jagiello—the grand prince of Lithuania—was elected king of Poland and merged the two countries into one state. By the middle of the seventeenth century, control of Byelorussian lands had passed to Poland. During the Great Northern War at the beginning of the eighteenth century (1700–1721), the Byelorussian and many Baltic lands were occupied by the armies of Charles XII of Sweden. In 1793 Byelorussia was annexed by Russia. By 1810 under Czar Nicholas I, the construction of the great fortress of Bobruisk was begun, which two years later successfully withstood a seige by Napoleon's troops.

But it was not only by migration and wars that the diverse ethnic backgrounds of the people of these lands were fashioned. For over two millennia trade had flourished from Scandinavia across the Baltic Sea, sometimes to Riga and down the Dvina (Daugava) River past Vitebsk to Smolensk (or, alternatively, overland to Smolensk over the routes controlled by the great city of Novgorod) and thence down the Dnieper to Kiev, past Ekaterinoslav to the Black Sea, and across to Constantinople or by sea route to the Aegean and Greece. Over these waters Greek, Persian, and Armenian traders linked the Greek world to the north of Russia and Scandinavia.

At the time that Hannah and Shlomo lived in Bobruisk in the 1860s, the city had about 24,000 inhabitants, perhaps half of them Jewish. The city built in the midst of forests held a commanding view of the Berezina River, a branch of the Dnieper. Rail now linked the town to Vilna (Vilnius), the great center of Jewish spiritual and cultural life scarcely 200 miles to the northwest, and to Ekaterinoslav (now Dnieperpetrovsk) toward the South in the Ukraine.

The raw facts of Hannah's life are straightforward. She and Shlomo at some point moved to Ekaterinoslav. In middle adult life, Hannah developed difficulty with her recent memory and eventually with even her personal care. A granddaughter remembers that

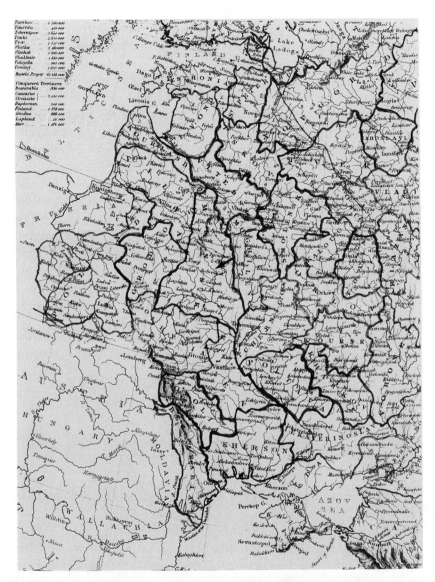

A map from 1853 showing "Russia in Europe" and Bobruisk, *(see arrow)* where Hannah and Shlomo then lived. In the text I have used the modern-day names for the various countries rather than those that appear on this map—for example, Latvia was then known as Livonia. (This map was prepared by the "Society for the Diffusion of Useful Knowledge," London 1853. Courtesy of the Harvard Map Collection, Harvard University.)

her own mother, Nadya, as one of Hannah's middle children, had to assist Hannah in combing her hair. Nadya never knew a normal mother. Nadya grew up in a family where all the children shared in Hannah's care. I have wondered what they must have thought about the cause of their mother's premature senility. But they had even more immediate concerns. After the assassination of Alexander II in 1881, in which a young Jewish woman played only a secondary role in the revolutionary conspiracy, the repression of the Jews of Russia became particularly intense, at times savage. The "Black Hundred," a strong nationalist and virulently anti-Semitic group, instigated pogroms across the vast breadth of Russia.

In the late-1890s a pogrom was launched against Hannah and Shlomo's community. During the onslaught, Hannah suffered an asthmatic attack—whether or not induced by fright no one can know—and died that evening. During the height of the pogrom Shlomo and the older children secretly buried Hannah under the cover of midnight to prevent the desecration of her body.

Thus it was that Hannah died with a disease of the brain which then had no name, and she was mourned in the darkness of night by her grieving family. Shlomo and Hannah begat

Chaim in 1869
Rosa in 1870
Elena in 1875
Morris in 1879
Ida in 1880
Nadya in 1883
Simon in 1885
Selma in 1887
Isaac in 1888

From these came the nine tribes of Hannah and Shlomo. Jeff, my patient, was a grandson of Ida.

We now know that the dementia that struck Hannah was genetic in origin. One hundred and fifty years ago no one would have guessed that the union of Hannah and Shlomo would bring forth hundreds of descendants who would spread across the Old and New Worlds. Many of these descendants, like Jeff, would be stricken with Hannah's disease, and some of the branches of the family would experience what great-grandson Ben, now in his early eighties, has aptly called "a personal biological Holocaust." And surely no one back then could have predicted that solving the riddle of the genetic defect in some of Hannah's progeny would become a major objective for many of the finest scientists around the world in the last decade of the twentieth century. For in discovering the nature of the genetic defects in the less common early-onset form of familial Alzheimer's disease, scientists may perhaps also find the clues to the enigma of the far more frequent late-onset form of the disease which some have called "the disease of the century."

And surely even if all of this could have been prophesied to the young Shlomo and Hannah, their belief would have been strained beyond all credulity had they been further told that any hope of deliverance for their progeny lay in the work of a young Augustinian monk beginning to study the propagation of traits characterizing the common sweet pea in a monastery garden.

No one would have believed such a prophecy, but so it would come to pass.

Gregor Mendel

Since humans first began to domesticate animals and cultivate plants, they realized that desired traits could be passed on to successive generations with increased frequency by selecting favored animals to breed or seeds to plant. Efforts to improve the quality of sheep for fine wool and of cattle for meat production were already well established in England and on the European continent by the end of the eighteenth century. In 1806 Christian Carl André proposed to the Agricultural Society of Brno, Moravia, a program for improved agricultural and industrial production based on extending the advances of natural science into these areas.[1] He urged the Society to establish a broadly based research program and accept the fact that it would be many generations before the results of the research would be of benefit to humankind.

At this time breeders realized that certain factors must influence heredity, but it had always been assumed that the determinants from the two parents *blended* to produce new determinants in the offspring. Yet this explanation failed to explain the complete reemergence in later generations of traits that had been apparently lost or blended in an earlier generation. F. C. Napp, the abbot of the monastery at Brno, had an intense interest in selective breeding and in 1836 emphasized the compelling need for a research program to determine the physiological basis of factors that determine heredity.

It was against this background that on August 9, 1843, in the year before Hannah's birth, Johann Mendel was admitted as a novice to the Augustinian monastery at Altbrunn on the outskirts of Brünn (Brno), the capital of Moravia, presently part of Czechoslovakia. He was born into a poor Moravian peasant family and required private tutoring to acquire a secondary education. After attending a theological seminary, Gregor Mendel, as he was now called, was encouraged by Abbot Napp to study the natural sciences at Vienna University. Upon Mendel's return to the monastery, he became a teacher of physics. But Mendel would soon turn his interests toward the problems of heredity posed by the abbot.

With the reforms wrested from the Habsburg monarchy following the revolutions of 1848, and a contemporaneous cultural shift in favor of increased scientific investigation, Abbot Napp gained a leading position in the Provincial Assembly and advanced the cause of natural science throughout the country and in his monastery. The extraordinary story of how a monastery in mid-nineteenth-century Moravia became a center of scientific research and served as the cradle for the experiments of Gregor Mendel has been researched by Vitezslav Orel.[2] For many years prior to Orel's historical research, it had been erroneously assumed that Mendel's work arose *de novo* from the mind of a solitary genius.

Mendel chose as his subject the common garden pea. He realized that clear-cut differences existed in some easily recognizable characteristics of the plant. For example, the ripe seeds could be either smooth or wrinkled, the unripe pods could be either green or yellow, and the stems could be either long or short. Mendel selected seven such pairs of characteristics to study. He first selected strains of plants that bred true for the various characteristics. Then he crossed such true breeding plants and studied the distribution of traits in their offspring.[3]

When he crossed plants having smooth round seeds with plants having wrinkled seeds, he found that *all* the seeds of the first new generation were round. Yet when he crossed plants from this first generation, he found that plants having either round or wrin-

Gregor Mendel standing second from right with members of the Augustinian Monastery at Altbrunn, Moravia, in the early 1860s.

kled seeds—but not both on the same plant—appeared within the second generation. But only one-fourth of the new plants bore wrinkled seeds; the ratio of plants with round to wrinkled seeds was very close to 3:1. The same pattern of inheritance was found for the other six traits studied. Mendel called the member of the pair that showed up in the first generation the "dominant trait." The trait that disappeared in the first generation only to reappear in the second was called "recessive."

Mendel then proposed that inherited traits are transmitted from parents to offspring by means of independently inherited "factors," which later, in 1903, would become known as "genes." The alternate forms of the genes which determine alternate physical characteristics would be called "alleles." For example, the alternate forms for brown or blue eyes are alleles. Mendel's first law, the "segregation of alleles," made it clear that alleles do not blend but remain independent of each other as they pass from generation to generation.[4] When both alleles are identical, the organism is said to be "homozygous" for that trait. If the alleles are different, the organism is "heterozygous." Fortunately for Mendel, the alleles of the traits he studied showed complete dominance, although this need not always be the case.

In his second law Mendel proposed that the factors or genes for different traits were assorted independently, according to the laws of chance: "the assortment of one gene does not influence the assortment of another."[5] Mendel was lucky a second time. The traits he studied were indeed transmitted independently as their alleles were located on different chromosomes. But chromosomes had not yet been discovered. Later it would be discovered that alleles close together on a given chromosome are not transmitted independently. Had it been otherwise, the great discoveries that have made the latter part of this story possible could not have occurred.

Although Mendel's work was published widely, it received little attention during his life. Perhaps this was because no physical basis for the site of his postulated factors was then known. Later, Mendel became abbot of the monastery and became immersed in administrative matters. When he died in 1884, Catholics, Protestants, and Jews alike joined his funeral procession.[6] No one then knew that they were also honoring the founder of a new science soon to be called genetics.

In 1900 Hugo DeVries in Amsterdam, Carl Correns in Tübingen, and Erich von Tschermak in Vienna confirmed Mendel's results and acknowledged the precedence of his discovery decades before. A fledgling science looked about for its next step.

Alois Alzheimer

Alois Alzheimer, the son of a notary of French ancestry, was born in Marktbreit-am-Main, Bavaria, in 1864, one year before Mendel presented the results of his experiments with plant hybrids to the Natural Science Society of Brno and five years before the birth of Hannah's firstborn Chaim. Alzheimer (1864–1915), together with Franz Nissl (1860–1918) and Emil Kraepelin (1856–1926), would make the decisive early contributions that made possible the systematic investigation of numerous brain disorders, including the one that afflicted Hannah.

Alzheimer attended medical schools at the Universities of Berlin, Würzburg, and Tübingen from 1882 to 1887.[1] Following his internship, he planned a career in clinical medicine, and he found a position as a resident physician at the city asylum for the insane in Frankfurt-am-Main in 1888. Nissl, four years older than Alzheimer, had recently become the attending physician of Alzheimer's department.[2] By this time, the legendary founders of neuroanatomy, Camillo Golgi in Italy and Santiago Ramon y Cajal in Spain, had already begun their pioneering work on the cell types (Golgi) and pathways (Cajal) of the normal brain. Nissl's earliest work led to the use of alcohol as a fixative for brain tissue and aniline dyes for staining the tissue. Nissl hoped that these technical advances would lead to the discovery of the anatomic correlates of psychiatric illnesses in the brains of his patients; hence, his decision to take a position at a mental hospital. Nissl's research objectives won the enthusiasm of his younger colleague[3] and thus began a scientific collaboration and close personal relationship between Alzheimer and Nissl that would last their lifetimes. F. H. Lewey, a colleague and biographer of Alzheimer, wrote of them:

> It was so perfect a companionship that it is impossible to decide which of the two owed more to the other. Nissl obtained his perspective from the laboratory bench, sitting there theorizing; Alzheimer's came from the clinic. Having known both of them, I would guess that the flood of startling ideas was Nissl's, but that it was Alzheimer who demonstrated their correctness histologically. Alzheimer had such a gift for describing what both had seen under the microscope, that the importance of their findings became immediately evident.[4]

Alzheimer and Nissl's lives would intertwine with that of yet another giant, Emil Kraepelin, "the Linnaeus of psychiatry" and one of the founders of biologically based psychiatry. When Kraepelin began his research, many psychiatrists regarded the dementias of middle life as psychiatric disorders of the mind in contradistinction to organic (i.e., pathological or biochemical) diseases of the brain. Thus, psychiatrists undertook many early investigations of dementia. Only much later did neurology as we know it emerge as an independent science and assume responsibility for research in this field in most, but not all, countries. Many leading psychiatrists of the late nineteenth century doubted the value of organic studies for any psychiatric disorder. For a time Kraepelin, too, shared this doubt. As Lewey wrote,

> Kraepelin's conversion to patho-anatomical research occurred rather late in his career, after the turn of the century. Before that it was different. Oskar Vogt relates that he visited Kraepelin at Heidelberg in 1894, bringing him greetings from Forel. Kraepelin asked what he planned to do in the future. "Brain anatomy of the psychoses," replied Vogt. "Then I must give you a bad prognosis," said Kraepelin, "for anatomy can contribute nothing to psychiatry."[5]

As a consequence of Vogt's visit, however, Kraepelin reassessed the relevance to psychiatry of Vogt's recent discoveries and those of Korbinian Brodmann, another renowned neuroanatomist. These two pioneers had observed that different regions of the cerebral cortex were anatomically distinct from each other with respect to the number and thickness of "layers" in each cortical region and that various distinctive cell types were found within each layer. The specificity in structure implied the specificity of function, and so perhaps structural or biochemical abnormalities in particular brain regions could lead to various expressions of psychopathology. Thus, Kraepelin finally became convinced that investigations of the brain itself were important for understanding the causes of the psychoses.

Kraepelin had already invited Nissl to his Institute in Heidelberg in 1895, although Alzheimer had remained in Frankfurt, becoming an attending physician. Still an avid clinician, Alzheimer had no desire to pursue a full-time research career. Only after he failed to be named the director of a mental hospital did he accept Kraepelin's offer to join him and Nissl at Heidelberg.[5]

Within the year Kraepelin moved to a highly prestigious academic position in Munich. Alzheimer followed Kraepelin to Munich and developed the Anatomic Laboratory of the Psychiatry and Neurology Clinic. Nissl remained in Heidelberg. Lewey describes the years from 1903 to 1908 as Alzheimer's finest: "He saw with clarity the determining factors responsible for the clinical ensemble. Patiently, he observed the most characteristic cases of each disease group, waiting sometimes for years before reporting his combined clinical and pathological finding."[6]

Alzheimer while still in his early forties published his classical descriptions of arteriosclerosis of the brain and senility. He astutely surmised that arteriosclerosis, caused by the accumulation of substances in the arteries to the brain, produced changes in the brain that were unrelated to the senile or aging processes observed in the brain in late life. He also provided the most detailed description of the pathological changes in the brain found in "general paresis," a condition that is characterized by psychosis, dementia, and paralysis and that usually develops many years after syphilitic infection of the brain. Syphilis was then very prevalent in Europe and untreatable until the discovery of salvarsan by Paul Ehrlich and his successful use of this "magic bullet" against syphilis microorganisms in 1910.

In Munich, Alzheimer became a great teacher, another role for which he is remembered today. Again in the words of Lewey,

> The twenty seats in his laboratory in Munich were always filled by students from all over the world. None of them could forget the many hours which Alzheimer spent with each individually, his large

head bent over the microscope, his pince-nez dangling on a long
string. The indispensable cigar he forgot as soon as he sat down, only
to light another as he moved to the next student; by the end of the
day some twenty big stumps were found around the laboratory. He
had little to beckon him home on time, as his wife had died.[7]

In the worlds of neurology, psychiatry, and neuropathology, Alois
Alzheimer is remembered for his extraordinary contributions to
describing brain pathology in arteriosclerosis, senility, and general paresis and for training and encouraging his protégés who
would lead the next generation of neuropathologists. Thus, it is
ironic that in the last quarter of the twentieth century his last
name has become a household word the world over for his brief oral
presentation of a single clinical case before the Meeting of South-
West Germany Psychiatrists in Tübingen in 1906. The case had
come to his attention quite by chance, and he published the contents of his paper in an apparently insignificant three-page note in
1907.

The patient was a 51-year-old woman who experienced an
unrelenting impairment of memory (amnesia). Her deficit in recent
memory was especially severe. I cite now from Alzheimer's classic
note in English translation[8]: "When the doctor showed her some
objects, she first gave the right name for each, but immediately
afterwards she had already forgotten everything." Her language
skills were also deteriorated. She spoke without intonation. "In her
conversation, she often used confused phrases, single paraphasic
expressions (milk-jug instead of cup), sometimes she would stop
talking completely. She evidently did not understand many questions." These symptoms point to what we nowadays would call
aphasia, or an inability to express speech or comprehend language.
But the case was more complicated than this: "She did not remember the use of particular objects." Neurologists call the inability to
carry out a previously learned skilled action in the absence of a
motor weakness or a sensory abnormality an "apraxia," and the
inability to recognize or identify a familiar object an "agnosia."

The patient died four and a half years after the onset of her illness. In the end she had become completely apathetic; she was
confined to bed where she lay in a fetal position, and she was
incontinent. The autopsy showed an evenly affected atrophic (i.e.,
shrunken) brain. Upon microscopic examination of the brain,
Alzheimer noted that numerous neurons, especially in the upper
cell layers, had totally disappeared. He then made an absolutely
new and startling discovery:

The Bielschowsky silver preparation showed very characteristic changes in the neurofibrils. However, inside an apparently normal-looking cell, one or more single fibres could be observed that became prominent through their striking thickness and specific impregnability. At a more advanced stage, many fibrils arranged in parallel showed the same changes. Then they accumulated forming dense bundles and gradually advanced to the surface of the cell. Eventually the nucleus and cytoplasm disappeared, and only a tangled bundle of fibrils indicated the site where once the neuron had been located.[9]

These bundles of fibrils are now referred to as Alzheimer's neurofibrillary tangles (see page 27). But Alzheimer noted a second major abnormality: "Dispersed over the entire cortex, and in large numbers, especially in the upper layers, miliary foci could be found which represented the sites of deposition of a peculiar substance in the cerebral cortex." These "miliary" or multiple foci were identical to the "plaques" which Emil Redlich had described as the hallmark of senile dementia as early as 1898. The presence of the plaques (see page 27) in *presenile* dementia (a dementia with onset at age 55 or younger) was especially significant. It would be some time before Alzheimer would discover that some cases of presenile dementia are characterized by plaques but not neurofibrillary tangles.[10]

It is perhaps rare that a single observable phenomenon, such as the deposition of a peculiar substance in a senile plaque, would evoke such intense and long-standing controversy regarding the etiology of a fundamental disease process. Within the course of only several years, Alzheimer and his contemporaries staked out opposing viewpoints on the crucial question about which comes first—the deposition of the peculiar substance or the primary destruction of brain cells. The early workers recognized that resolution of this issue might be decisive in understanding the cause of Alzheimer's disease, senile dementia, and perhaps even the milder yet universal declines in mental function that occur in so-called normal aging.

Redlich considered the plaques he had described as "glial proliferations," an abnormality of the nonneuronal brain cells. He thought the plaque corresponded to an extensively proliferated glial cell that had lost its nucleus and replaced a necrotic (dead) neuron. Yet Oskar Fischer, who confirmed Redlich's description of the presence and distribution of the plaques in senile dementia in 1907, contested Redlich's hypothesis as to their etiology.[11] Fischer, as had Alzheimer, thought that the plaques originated in the accumulation of a peculiar foreign substance in the brain.

From left to right Alois Alzheimer, Emil Kraepelin, Robert Gaupp, and Franz Nissl. The four are photographed together (probably around 1908–1910) on an excursion boat on the Starnberger Sea. The photograph was sent to me by Dr. Paul Hoff. (Courtesy of the Psychiatric Hospital of the Ludwig—Maximilians—University, Munich.)

Francesco Bonfiglio (1908) disagreed with Fischer.[12] After careful study of the evolution of the plaques, Bonfiglio believed he could exclude any accumulation of a foreign substance as a primary etiological factor. Instead, he believed that the abnormality began inside a neuron and in the nerve terminals that surround it.

Although the early workers contested the initial causal event, they agreed that the unknown process that can lead to the production of a plaque can occur in virtually everyone given a long enough life span. When they compared normal aging versus senile dementia, they observed differences in the number, size of the plaques and especially in the regions of the brain where they occurred; yet the plaques in the two cases were essentially indistinguishable. As for clinical symptoms of cognitive and memory decline, by 1910 Kraepelin had recognized that some decline inevitably occurs in every very elderly person. He found it impossible to set a boundary between the decline of normal aging and the first signs of senile dementia.

In 1910 Kraepelin proposed naming the presenile form of dementia after Alois Alzheimer. He wrote:

Photomicrograph of the "hippocampus," a part of the brain important for the registration and recall of new memories. Here a 69-year-old patient with "sporadic" Alzheimer's disease shows the classical neuropathological features of the disease: senile plaques (longer arrow) and the neurofibrillary tangles (smaller arrow). A modified Bielschowsky stain was used, and the magnification is 80X the original size. (Courtesy of Dr. Carol F. Lippa, Department of Neurology, University of Massachusetts Medical Center.)

> The clinical interpretation of this Alzheimer's disease is still con- fused. While the anatomical findings suggest that we are dealing with a particularly serious form of senile dementia, the fact that this disease sometimes starts already around the age of 40 does not allow this supposition. In such cases we should at least assume a "senium precox," if not perhaps a more or less age—independent unique dis- ease process.[13]

Alzheimer published the results of his second case of presenile dementia in 1911. He accepted Kraepelin's distinction between "presenile dementia"—with its onset before ages 50 to 55, its accompanying psychiatric symptoms, and its early development of amnesia, aphasia, apraxia, and agnosia which have a rapidly pro- gressive course—and the late-onset "senile dementia." He even believed that gross anatomical distinctions existed between the two entities. For example, the pathology of the frontal lobes was more severe in senile dementia than in presenile dementia, which most severely affected the temporal and parietal lobes. The temporal lobe is involved in certain aspects of memory and several important speech functions; the parietal lobe subserves many functions, including encoding our body image and its relationship to other objects in visual space; and the frontal lobe functions include foresight, initiative, and several aspects of ideation.

Alzheimer was never certain as to whether or not presenile disease was simply an unusually severe and premature form of senile dementia.[14] As Gaetano Perusini wrote the same year (1911), "In any case, even though, as Alzheimer believes, these morbid forms (i.e., Alzheimer's disease) do not represent anything but atypical forms of senile dementia; at times, their extraordinary early onset and the features of their clinical course are sufficient to consider them a separate group."[15]

In the years between 1906 and 1911, Alzheimer entrusted four brains with senile dementia to Bonfiglia and Perusini. By 1910 Perusini doubted that the plaques arose from dying cortical nerve cells and denied that the plaques were specific to the dementias. A colleague had even found "formations" that resembled plaques in the cerebral cortex of an aged pig! Another colleague, Dr. Ugo Cerletti, had sometimes found both plaques and even tangles in the brains of normal old people. By 1911 Perusini summarized his conclusions, affirming that he could find no clear-cut distinctions between senile dementia and normal aging at least as far as the presence—though not the extent—of the senile plaques was concerned.

The early masters had gone as far as they could go with their stains and light microscopy. The wonder of it is how far they had carried the problem in the five years from 1906 to 1911. Many years would pass before anyone would go beyond the work of Alzheimer and Perusini. For a time, with much justification, presenile dementia was called Alzheimer–Perusini disease, especially by workers in Italy.

In 1912 Alois Alzheimer accepted the position of professor of psychiatry and director of the Psychiatric Institute at the University of Breslau. On the train to Breslau he became ill from a heart condition and had to remain in the hospital for a time before he could assume his responsibilities at the university. With the onset of World War I in 1914, he lost his research assistants. His own workload increased even as his own health deteriorated. Dr. Robert Gaupp from Tübingen visited him in Breslau in the fall of 1915 and reminded him of the international group of Americans, Italians,

Dutch, Rumanians, French, and Swiss physicians who had worked with him in Munich. According to Gaupp,[16] Alzheimer smiled in his quiet way and said, "All those will also come back after the war. I do not think that they are going to avoid me as a barbarian because they did not have to complain about me." But this was not to be. In the fall of 1915 his rheumatic endocarditis recurred. His heart and kidneys were failing. He quietly told Gaupp that he must now prepare to ask his sisters and his brother-in-law to take care of his children. Shortly afterward, Alois Alzheimer died on December 19, 1915 at the age of 51, ironically, the same age at which his famous patient had developed her first symptoms of the disease that now bore his name.

Alzheimer's life-long colleague, Franz Nissl, died on August 11, 1918 at the age of 58. It was left for Gaupp, who had witnessed the age of the giants, to write: "What I wrote three and one-half years ago at the death of Alzheimer, I should like to repeat today at the grave of his and our dear friend, Nissl. Our life is poorer and it became colder around us when we saw him depart. We have loved him too much to forget him."[17]

PART II

Into the Wilderness

The Descendants of Hannah and Shlomo

From 1903 to 1905 new waves of pogroms once again spread over Russian lands. The memories of Hannah and her death during an earlier pogrom in the 1890s were not forgotten by her threatened family. But now Theodore Roosevelt was president of the United States, and the gates of America were open to the downtrodden of Eastern Europe. Beginning in 1903, eight of the nine sons and daughters of Shlomo and Hannah eventually immigrated to the United States. Even the widower Shlomo, then over 60, joined the exodus. The family had escaped the persecutions, pogroms, and poverty of Czarist Russia, but for some there would be no escape from a threat hidden within them.

Arriving in America, Hannah's sons and daughters followed paths typical for immigrants of that period. A daughter remained in the East and worked in the sweatshops of the garment industry; several of the older children settled in railway towns along the Mississippi; and still others pressed across the country to California, all the while remaining in close contact with each other.

Great-grandson Ben, who escaped the defective gene, shared a photograph with me of Shlomo and 29 of his American grandchildren and great-grandchildren taken in 1918. Shlomo, at 80, stands tall, erect, fully bearded and wearing the traditional skull cap of Orthodox Jews. His Americanized descendants surround the patriarch in row upon row according to age. Another grandson, not in the photograph, would soon return unscathed after serving in the British Army in World War I. There is a measure of pride in Shlomo's eyes, but his expression seems tempered by a discomforting apprehension. Could he have feared, even then, that some of his older children were in the earliest stages of the disease that had claimed Hannah? Ben knew about the disease as early as 1923 (the year after Shlomo had died) when as a boy of 11 he saw his mother's aunt, who was only in her late forties, sitting in a wheelchair and being fed.

Ben also remembers when Chaim, his grandfather and Hannah's firstborn, became afflicted by Hannah's disease. In that instant, Ben at once experienced his sorrow for Chaim, his anguish that his own mother was now marked as at risk, and his own terror that he too might in time suffer the fate that had claimed Hannah and now Chaim. Ben recalls the years that followed:

> My grandfather was a very pious man. My memories of him included many images of him being chastened by his wife for his forgetfulness. She kept repeating her injunctions while in tears. It appeared that the only living thing that tolerated his existence was his faithful terrier Belle. When my grandfather died in 1927 from a malignancy, it must have been a relief to both him and God, who was always in his prayers. Perhaps God, after all, rescued him from five to ten years of a prolonged vegetative existence.

Ben was a medical student in 1935 and elected a clinical rotation in psychiatry. His supervising physician, who turned out to be another of his cousins, had just hospitalized his own father, a pleasant but demented man of 55, on whom a clinical diagnosis of Alzheimer's disease had been made earlier. The patient, the fourth child of Hannah and Shlomo, died in 1936. An autopsy ordered by the patient's physician-son confirmed the diagnosis. Ben now realized for the first time that Hannah's disease was a familial form of Alzheimer's disease. At this time there were only several scattered reports in the medical literature that the disease could be familial, and these reports are now known largely in retrospect. They made little impact at the time, but Ben knew.

That same year Ben's mother, Anne, then 48, diagnosed herself as having the earliest signs of the disease. Her illness progressed rapidly. Ben recalls the 1930s:

> These were very difficult years. Day and night, week in and out, the merciless days passed, each filled with a thousand distractions and heartaches. All we had was a diagnosis and tears. There was no answer to the unknown plague about which we knew nothing, could read nothing, and only knew that the disease had been described in about 1906 by a doctor in Munich, Germany. These were days without hope. Where did it come from? How did it get started? Yet these were modern times. Why did no one know more about the disease? It was hard to believe that in this whole world our family was the only one possessing this taint. Would it ever end?

And this isolation imposed its own toll. Ben remembers, "We never discussed the condition with strangers. We had very little discussion among ourselves about Alzheimer's disease. It was almost that if you did not discuss it, that it would go away. So horrible in our minds were the implications of this gene that we felt by discussing it, we gave it substance." His terror of the disease, suppressed by the distracting activities of the day, intruded in fearful dreams and inconsolable anguish by night.

As Ben's own medical studies continued, he himself became interested in studying the nervous system. He learned to recognize the signs and symptoms of neurological disorders including the one that afflicted Hannah, Chaim, and now Anne. His days on the neurology service could no longer be free of the spectre of working with demented patients. Ben could no longer contemplate a career in neurology, "as then my days would be as difficult as my nights."

As the 1940s began, Ben's mother required daily care. "All personal plans had to be modified. Arrangements had to be made for a caretaker," Ben recalls. "Somebody would have to make the sacrifice. The only sister? Which brother?" This problem reached a crisis by the time World War II began. Ben and his brothers joined the Army, and in 1942 Ben was sent overseas. When he returned in 1945, his mother would no longer recognize him.

Meanwhile, another of Ben's great aunts had developed the disease by the early 1930s, but no diagnosis was then available. This woman, Ida, a daughter of Hannah, bore seven children; two had died in early childhood. But still in the prime of her life, at 22 Ida bore Jared, at 26 Larry, at 30 Sarah, and at 34 Helen. And then at 39 she bore her last child, a son, Charles. Ida's husband died in 1938, and Ida died a year later. An autopsy, the second on a member of the family, was done and confirmed the diagnosis of Alzheimer's disease. If anyone in the family had harbored doubts as to the nature of the hereditary illness, there could be none now. Ida left five children, the youngest, Charles, only 18, went to live with Jared, his oldest brother. Charles, although a grandson of Hannah, was eight years younger than great-grandson Ben, a descendant of Hannah's oldest son (see facing page).

By the time Charles was in medical school. Jared, who had been a brilliant physician, developed the disease. By the mid-1950s, Larry and Sarah had also developed early signs of the illness. Charles was now 35. Bryna, a granddaughter of Hannah and a first cousin of Charles, who like Ben escaped the abnormal gene, recounted the early history to me. She became the family historian, gathering records on all family members, while Charles, now a pathologist, desperately tried to interpret the data.

Bryna told me that when the disease had first struck members of the second generation, the family members nurtured the hope that the virulence of the disease would somehow attenuate in future generations, for they assumed that whatever factor caused the disease would blend with new genes coming into the family. Interestingly, they intuitively made the same assumptions regarding a blending of "factors" that pre-Mendelian breeders had made earlier. But Charles recognized that neither the severity of the disease nor its age of onset had changed in the three generations. Whatever caused the disease was as immutable as the genes that Gregor Mendel had described in his experiments on sweet peas. But Charles was not working with sweet peas; he was studying his own ancestors and loved ones. And if Gregor Mendel could breed a new crop of sweet peas in several months, Charles had to wait years and decades for enough data to accumulate to draw accurate conclusions.

During these decades many members of the third generation feared that all of them would develop the disease if they lived long enough. Bryna did not sense that she would escape the disease until her own mother was still free of it at age 60.

As Charles studied the data he found that the age of onset of the disease ranged from 42 to 54, with a mean age of onset of 46 to 47. But he knew these data were based on relatively small numbers and that most biological phenomena followed a bell-shaped curve. So it was possible that a few people destined to develop the disease might become symptomatic a bit earlier or a bit later.

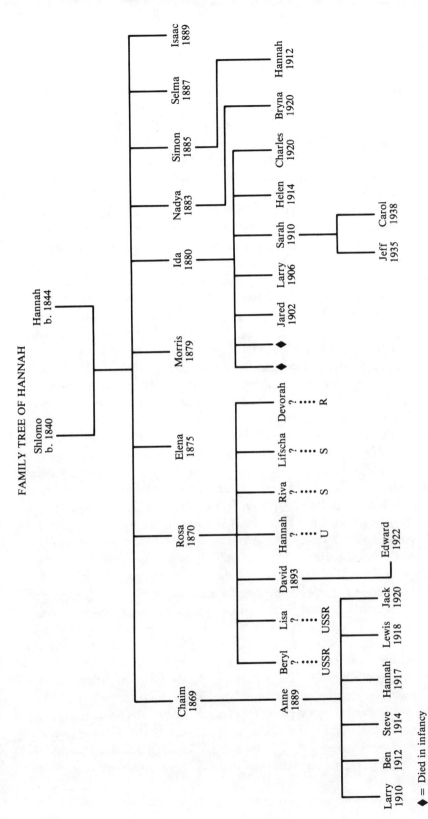

FAMILY TREE OF HANNAH

= Died in infancy

Diagram showing the family relationships between Ben, Charles, Bryna, and other descendants of Hannah and Shlomo.

In time, Charles calculated that four of the eight American members of the second generation had developed the disease, and of those who did so, about one-half of their children in the third generation also inherited the gene. He was now able to conclude that the factor for the disease was a dominant gene. The gene was not sex-linked because transmission occurred from mother to daughter as well as from mother to son. Transmission also occurred from father to son. Thus, "mitochondrial" transmission, which occurs almost entirely through the mother, was also essentially excluded, but Charles could not have known about this form of genetic transmission then. Charles correctly surmised that the gene was an autosomal dominant gene—that is, one that must reside on one of the 22 autosomal (i.e., non-sex-linked) pairs of chromosomes. And since fully 50 percent of the at-risk family members eventually developed the disease, the disease was fully penetrant (i.e., the disease would eventually develop in every individual who inherited the genetic defect).

In those branches of the family where Hannah's descendants have survived past the age of 54 without developing symptoms of the disease, no subsequent cases of familial Alzheimer's disease have occurred even to this day. Once an at-risk member has escaped the disease, his or her progeny have also escaped (unless, of course, an escapee marries another carrier). Thus it was that several lines from Hannah and Shlomo's union escaped the disease; their descendants today live without fear of this dread disease. Unfortunately, however, only three of Hannah's eight children who came to America escaped the abnormal gene. A son who was not known to have had the disease died at age 50. He left children who subsequently developed the illness. Later, an autopsy would prove that this illness, too, was Alzheimer's disease.

Hannah's second-born child, Rosa, who remained in Russia, bore seven children; this much was known because one of Rosa's sons eventually came to the United States. In late midlife this man developed another neurodegenerative disease called ALS or amyotrophic lateral sclerosis. Before his death, he gave Bryna the last known address, dating from the late 1930s, of a sister who had lived in Ekaterinoslav, now Dniepropetrovsk. But as Bryna compiled her records in 1956 she had no hope of contacting the Russian branch. Soviet troops patrolled the streets of Budapest, and a tense Cold War seemed a permanent fact of life. Nevertheless, she kept the address.

During the next few years, Charles's remaining sister also

developed Alzheimer's disease. Now Charles himself had entered the age of risk even as he had so defined it. The fact that Charles's four siblings had developed the disease neither increased nor decreased his own risk. That risk was still 50 percent unless by some remote possibility, his father had also carried the same abnormal gene, in which case Charles's risk would be 75 percent. Even more remotely, if his mother had received two copies of the defective gene—one copy from each of her parents—Charles's risk would be 100 percent.

Charles remained driven to find the cause of Alzheimer's disease. He had earlier followed in the footsteps of Gregor Mendel, and now he followed in the footsteps of Alois Alzheimer. Well trained in neuropathology as well as general pathology, Charles carefully studied the slides of the brain pathology on various members of the family, even those of his own brothers and sisters. He was determined that their deaths would not be in vain. Wherever possible, something more had to be learned from each case.

If Charles could not find the cause of the disease, he knew at least one way to stop its spread. It was a desperate choice, but Charles made it: he decided not to father a child. He would not risk spreading the disease, but it was up to each member of the family to make his or her own choice. Ben chose as Charles had. Later, Ben developed the attitude that if an at-risk individual desperately wanted to have a child, then he or she should have just one offspring. That way, even in the worst case, there would be no increase in the number of affected individuals in the next generation. Bryna chose to risk, having the one child.

By 1964 all four of Charles's brothers and sisters had succumbed to Alzheimer's disease. Only Charles, still at risk, was left.

Mendel's Heirs

Perhaps no one played a greater role in the development of classical Mendelian genetics than Thomas Hunt Morgan, who was born near Lexington, Kentucky, in 1866, the same year that Mendel's discoveries were published.[1] Thomas was born into one of America's most illustrious families, a family of statesmen, bankers, and soldiers. His uncle, Brigadier General John Hunt Morgan, the legendary "thunderbolt of the Confederacy"—a treacherous raider or a gallant, daring soldier, depending on one's political view of the Civil War—had been killed two years before Thomas was born.

In 1886 the 19-year-old Morgan traveled to Baltimore, the home of his mother's family, to enter a relatively new, then little-known school called Johns Hopkins University founded scarcely 10 years before. At the founding, Thomas Huxley, England's greatest biologist and leading exponent of Darwin's evolutionary theories, gave the inaugural address. From its start, the school's Biology Department had a strongly pro-Darwinian faculty. Chromosomes had only recently been discovered as components of cell nuclei in 1875, and within the span of only two years, scientists discovered that the fertilized egg receives half of its chromosomes from each parent. The central question of Morgan's life and work could now be phrased—namely, how did the constituent components of chromosomes determine the development of the embryo and the specificity and uniqueness of the resulting animal? Shortly after graduation from Johns Hopkins, Morgan joined the Bryn Mawr faculty and began research in embryology and regeneration.

In 1904 Morgan moved to Columbia University where he became the first professor of experimental zoology. It was at this time that two new ideas would profoundly change his life and the future course of genetics. First, in 1900 he visited Hugo DeVries in Holland, one of the three European scientists who had just rediscovered and confirmed Mendel's work. Second, from DeVries's theory Morgan learned that new species originated over time through a process known as mutation. Morgan had been philosophically opposed to the notion of constant struggle which Darwinian evolution demanded as the main factor in the development of new species. Concerning mutations he wrote, "From this point of view, the process of evolution appears in a more kindly light than when we imagine that success is only obtained through the destruction of all rivals."[2]

Morgan, like many of his contemporaries, was intrigued by J. B. Lamarck's thesis that acquired characteristics could be passed on from parent to offspring. In 1908 Morgan encouraged a graduate student to see if the eyes of fruit flies would atrophy if they were raised in the dark for numerous generations. Sixty-nine generations later, the experiment had failed, but Morgan seized on the fruit fly as his experimental animal for his life's work. *Drosophila*, the prolific, rapidly breeding fruit fly, would become the ideal experimental animal for experimental classical genetics.

Morgan, in yet another test of an idea proposed by DeVries, then tried to determine whether mutations in the flies could be produced by radiation. After two years of work with no success, Morgan was almost ready to give up when a single male drosophila was born with white rather than the customary red eyes. Morgan set about to breed this single fly. Ten days later it had produced 1,240 offspring; all but three had the red eyes expected of a Mendelian dominant trait. (In theory, all should have had red eyes; the discrepancy has never been explained to everyone's satisfaction.[3]) In the next generation, there were 3,470 descendants of the original white-eyed fly, and almost one-quarter had inherited its white eye color. This was the classic prediction and a confirmation of Mendel's model for the inheritance of a recessive characteristic!

Yet when Morgan looked closer, he discovered that the sex distribution of the white-eyed flies was not random. In fact, one-half of the males but none of the females had inherited the white eye. He immediately understood that the factor, or "gene," producing eye color and the factor producing the sex were inherited together or "linked." He then found that genes for mutants with eye colors other than white were "segregated" independently of the sex factor. In other words, these genes could not be on the X chromosome. By 1911 he had established that a single chromosome carried both the sex factor(s) for maleness and the gene for white eyes. Another factor determining a "miniature wing" was also linked to the sex factor on what Morgan called the "X chromosome." So now Morgan had proved that three distinct genes resided on a single chromosome. Then from a chance observation sprang a series of profound insights into the very inner working of the chromosome itself. Two characteristics, both of which were presumably due to sex-linked genes, were inherited together in most but not all offspring. Morgan's perhaps greatest discovery followed. As Shine and Wrobel wrote,

> Morgan then hit on an explanation for the failure of the two sex-linked genes to cosegregate together; perhaps they were not situated *close together* on the X chromosome. If during meiosis the two X chromosomes (like the other pairs of chromosomes) exchange their genes, well separated parts of the chromosome would be likely to exchange. For two genes close together, this was more unlikely.[4]

Morgan termed the tendency of genes to stay together as "linkage," and the process of interchange of genes between chromosomes during meiosis as "crossing over." (Meiosis refers to the gene-shuffling process during the formation of the sex cells, which results in each of the reproductive cells containing a single rather than the double set of chromosomes found in somatic cells.) Shine and Wrobel succinctly explained this fundamental idea of genetics.

> Genes are distinct units like cards, which are played separately and do not blend. Chromosome pairs come to lie intimately together during meiosis, at which time the maternal and paternal chromosomes are shuffled so that their genes become exchanged in the same way as cards of two decks of playing cards that are shuffled together. The closer any two cards lie together in a deck, the less likely a single cut will separate them. Indeed, there is only about a 2% chance of cutting between any two particular adjacent cards and a 98% of not doing so.[5]

Thomas Hunt Morgan at work at his laboratory at the
Woods Hole Marine Biological Laboratory. (Courtesy of
Morgan's youngest daughter, Isabel M. Mountain.)

This idea of the exchange of chromosomal material had been suggested by previous workers, but Morgan and his group proved the
theory and extended the significance of the observation. Morgan
had hired two teenage students at Columbia to help him identify
mutants in *Drosophila* and to count the segregation of traits after
selective matings. One of these students, Alfred H. Sturtevant, realized that the relative distance between genes and their order on a
given chromosome could be determined by the results of genetic
crossings. Sturtevant wrote,

> In the latter part of 1911 I suddenly realized that the variations in
> strength of linkage, already attributed by Morgan to differences in

the spatial separation of the genes, offered the possibility of determining sequences in the linear dimension of a chromosome. I went home and spent most of the night (to the neglect of my undergraduate homework) in producing the first chromosome map, which included the sex-linked genes y (yellow body color), w (white eye), v (vermilion eye), m (miniature wing), and r (rudimentary wing) in the order and approximately the relative spacings that they still appear on the standard maps.[6]

In time, Morgan's and Sturtevant's results would lead to the very principles used today to infer the position of unknown genes responsible for any number of genetic diseases, among them Alzheimer's. Later, Herman Muller, one of Morgan's finest students wrote: "Morgan's evidence for crossing over and his suggestion that genes further apart cross over more frequently struck as a thunderclap: hardly secondary to the discovery of Mendelism, which ushered in that storm that has given nourishment to all our modern genetics."[7]

In 1927, at the age of 61, Morgan was still at Columbia and was nearing the customary retirement age of 65. But he was not too old to be invited to organize a new division of biology at the California Institute of Technology, which by then already had an astonishing concentration of fine minds in physics and chemistry. Morgan accepted the challenge and set about to build a division of biology that would be receptive to the new ideas of biophysics and biochemistry and other emerging sciences. Morgan brought some of his team from Columbia with him, including Sturtevant and Calvin B. Bridges, another superb geneticist, who along with Sturtevant had begun work with Morgan as a teenager. Another highly productive period of research followed.

In 1933, on the hundredth anniversary of the birth of Alfred Nobel, Thomas Hunt Morgan was awarded the Nobel Prize for his work on the chromosomal theory of inheritance. He acknowledged the joint contributions of Sturtevant, Bridges, and Muller in the work of almost a generation. Yet Morgan was surprised he had received a Nobel award in the category of "Physiology or Medicine." He did not foresee that genetics had much to contribute to either physiology or medicine apart from genetic counseling.

Morgan had paid little attention to the work of the English physician, Archibald Garrod, who, as early as 1908, had realized that certain metabolic diseases in humans were hereditary traits. Garrod had called such diseases "inborn errors of metabolism" and hypothesized that they were due to the absence of specific enzymes, each of which was controlled by a specific gene. Yet it would be only a year after Morgan's award that phenylketonuria, the first metabolic disease that was eventually preventable, was discovered. In its early period of growth, genetics had split into two branches that scarcely interacted. In the one branch belonged workers like Morgan who inferred the relative locations of genes on specific chromosomes from their observation of expressed characteristics, and in the others were those, primarily trained as biochemists, who studied "enzymes" and their dramatic facilitory effects on biochemical reactions.

Later, in 1940, George Beadle, who had once been a postgraduate fellow at Cal Tech under Morgan, joined with the biochemist, Edward Tatum, to prove the one gene–one enzyme concept which in time would serve as a first principle for gene mappers and enzymologists alike. Still no one could fault Morgan in 1933 for doubting that knowledge of genetics could ever lead to the treatment or cure of a genetic disease. And as for knowing anything about the actual biochemical structure of a chromosome or gene, well, this seemed beyond comprehension, even beyond a dream.

Yet in time, a number of seemingly independent developments would give birth to the dream. Looking back, I have tried to recall some of the landmark theoretical concepts and experimental breakthroughs that led eventually to the epic discoveries on the structure of genes. Assuredly, this retrospective approach may serve as a convenience for an historian, but it can never presume to portray the actual process of scientific discovery itself, with its numerous fits and starts, its ferment and frustration, and its contending and often contentious claimants. For our story, it is difficult to find a single trend, but it was through the work of Morgan, Linus Pauling, Niels Bohr, and Max Delbrück that an essential unity of the physical and biological sciences would come to be appreciated and create new possibilities for future breakthroughs.

Morgan's hope for a coalescence of ideas from physics, chemistry, and biology came to be realized, in fact, at his own Cal Tech, but in ways that would have seemed unimaginable to him, a non-mathematically minded empiricist. Unbeknownst to him, Linus Pauling, one of the authentic intellectual giants of our century who was at Cal Tech during the same period, established the fundamental principles governing the nature of the chemical bond in the period from 1931 to 1935. However, only in the early 1950s would the full significance of Pauling's work for genetics come to light.

At about the time Morgan received his Nobel Prize in recognition of the founding of classical genetics and while Pauling was creating the very foundations for a future molecular genetics, another unique seed was being sown, this too inspired from the world of physics. In Copenhagen during the golden age of physics between the wars, Niels Bohr educated and encouraged two generations of many of the world's finest physicists. Bohr, like Albert Einstein, brought to science and his students a magnanimity of spirit and an openness in the search for the truth. Bohr ceaselessly practiced a Socratic approach to the analysis of great scientific problems. Max Delbrück, a young German physicist and himself a pioneer in the development of quantum mechanics, developed in the Bohr tradition, both in his approach to scientific problems and in his openness of spirit. According to Garland Allen,

> Bohr had emphasized (among other things) the vast number of important problems biology offered. Bohr had suggested to Delbrück that the simple reduction of biological problems to physics and chemistry (especially the atomistic type of physics and chemistry adhered to by most biologists) was inadequate to the solutions of critical problems such as those of heredity and development. Those imbued with the new ideas of quantum physics recognized that the act of breaking an organism down into its individual component parts destroyed the very organizational basis that ought itself to be the object of future study.[8]

Subsequently, Delbrück proposed some ideas "about the physical properties of the gene, in a youthful paper in 1935."[9] Two years later, he arrived at CalTech to study biology in part because of that institution's strengths in genetics established by Morgan. However, no sooner had he arrived than Delbrück became convinced that he could learn little more from Morgan's now classical approach to genetics. He realized that further fundamental discovery in genetics would have to rely on approaches at the molecular level. As important as Delbrück's prolific contributions to molecular genetics would become over the next half-century, it is hard to overestimate the impact his earlier "youthful paper" would have on subsequent developments in genetics after the physicist Erwin Schrödinger published a book *What Is Life?* based on Delbrück's ideas. As Horace Judson, who has documented the birth of modern molecular biology in his wonderfully readable book *The Eighth Day of Creation*, has written of Schrödinger's book:

> This was a short work, published in 1944, that speculated about the physical basis—the atomic and molecular basis—of biological phenomena. In it, Schrödinger, the founder of quantum mechanics in its

modern form, an Austrian of Roman Catholic background living in quiet exile in Dublin, popularized some suggestions about the nature of the gene, in the light of the principles of quantum mechanics, which had been made in Berlin in 1935 by another physicist, Max Delbrück.[10]

One prophetic idea raised the possibility that the information content of genes could be encoded by the sequence of molecules along an "aperiodic" crystal. Here for the first time was an idea of how an ordered sequence of different combinations of some type of yet unknown molecules strung out along a chromosome might comprise a gene. By 1938 another physicist, Pascual Jordan, proposed that biological molecules such as genes and antibodies replicate themselves according to his idea that groups of molecules would attract identical groups. In 1940 Linus Pauling and Delbrück pointed out the fallacies in Jordan's understanding of the nature of the chemical band (by then established by Pauling) and proposed instead a principle of complementarity for molecular bonding and replication. Biological systems would achieve maximal chemical stability when oppositely charged chemical groups mutually attracted. Thus, two molecules with complementary surfaces would fit together like "die and coin,"[11] and one such surface would serve as a template for the synthesis of its complement. Yet without any factual knowledge about the chemical nature of the genetic material, Pauling and Delbrück's ideas could not be carried further.

Morgan remained at his post at Cal Tech until 1942 when he retired. In his later years he returned to the study of embryology, his first love. Sturtevant then ran the division from 1942 to 1946 until George Beadle returned from Stanford. Even after retirement, Morgan continued his work. He died at 79 in 1945. The year before Morgan's death, Oswald Avery, together with Maclyn McCarty and Colin MacLeod, discovered that the peculiar "transforming factor" that conferred pathogenicity to the bacterium *diplococcus pneumoniae*, the bacillus that causes pneumonia, was composed of deoxyribonucleic acid—DNA. One wonders whether Morgan knew of Avery's discovery or appreciated, as few did at the time, its significance.

Morgan's life had spanned a remarkable period from the end of the American Civil War to the end of World War II, from the publication of Mendel's thesis about independent factors to the discovery that a genetic material, perhaps a gene, might be composed of nucleic acids rather than a protein structure, as had been previously believed by most scientists.

Still, the idea that chromosomes and genes could be composed of proteins rather than nucleic acids died only very slowly. James Watson states that "some skeptics preferred to believe that Avery and his colleagues had somehow missed seeing the genetic protein and that DNA was required for activity in their assay only because it functioned as an unspecific scaffold to which the real protein genes were fixed."[12] Avery's work had to be extended and generalized to other organisms before his conclusions were accepted.

Now it was possible to dream about knowing the structure of a chromosome, and perhaps no one was more driven by this dream than the precocious James Watson who completed his Ph.D. in genetics under Salvador Luria in 1950 at the age of 22! Watson knew the work of Avery and his successors and reasoned that the genetic material must be DNA itself. He was also convinced that Delbrück and Schrödinger's basic ideas had to be fundamentally correct—namely, that genetic information could be encoded by molecular sequences composing an aperiodic crystal.

The chemical composition of DNA was known. DNA was composed of differing amounts of only four nucleotides: "Each nucleotide contains a phosphate group, a sugar moiety, and either a purine or pyrimidine base (flat, ring-shaped molecules containing carbon and nitrogen."[13] A figure from a text co-authored by Watson[14] shows the two purines *adenine* (A) and *guanine* (G) and the two pyrimidines *cytosine* (C) and *thymine* (T) found in DNA. (The other pyrimidine *uracil* (U) appears in a second kind of nucleic acid—ribonucleic acid—or RNA; thymine appears only in DNA.) A second figure shows how the sugar moiety—deoxyribose—is attached to a phosphate group and one or another of the four DNA bases.

This much was known when Watson began his postdoctoral work in Denmark with Hermann Kalckar, who in the following spring of 1951 took Watson with him to Naples for several months' work. There Watson attended a lecture given by Maurice Wilkins from King's College London who showed a slide "of a new X-ray diffraction pattern Wilkins had obtained from what he said was a crystalline form of DNA." Watson had never heard of Wilkins and knew nothing at all about crystallography. But he saw the essential point: "if DNA could form crystals, it must have a repeating, regular, orderly structure, and so could be solved."[15] (The science of X-

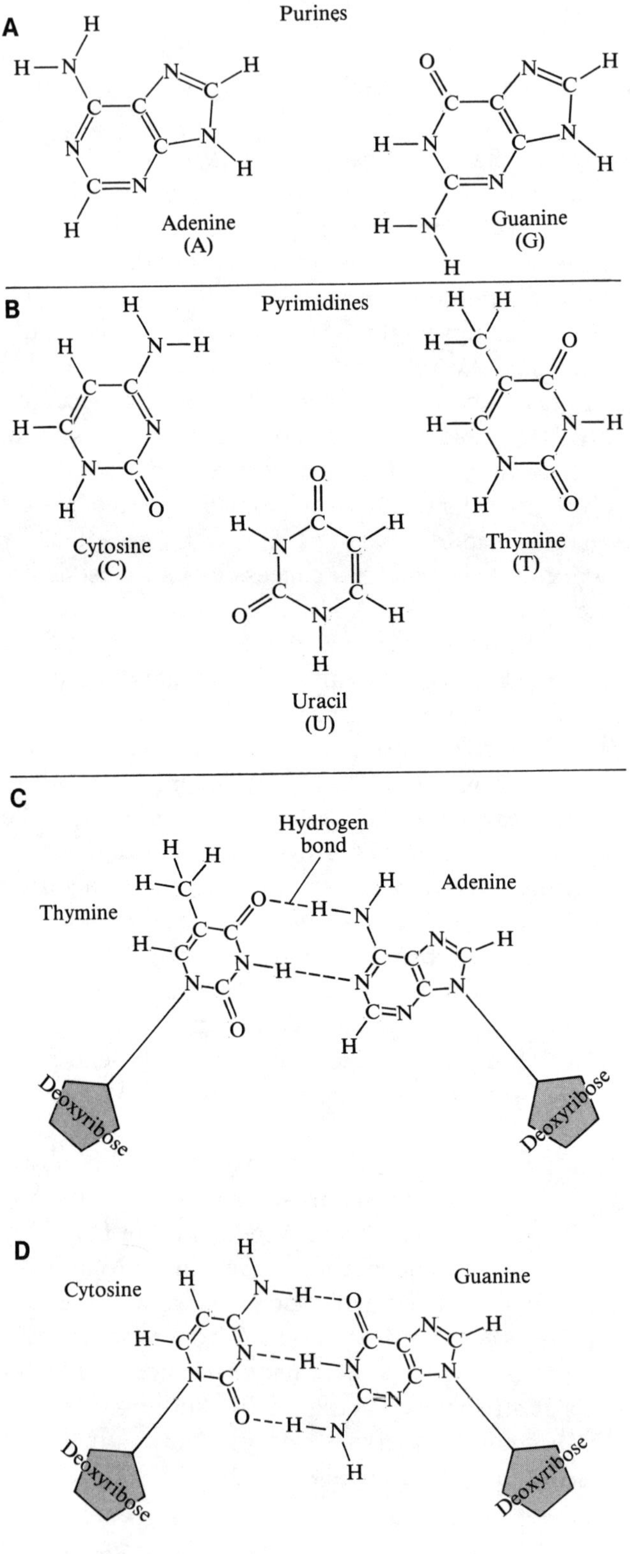

A Purines

Adenine
(A)

Guanine
(G)

B Pyrimidines

Cytosine
(C)

Uracil
(U)

Thymine
(T)

C

Hydrogen
bond

Thymine

Adenine

Deoxyribose

Deoxyribose

D

Cytosine

Guanine

Deoxyribose

Deoxyribose

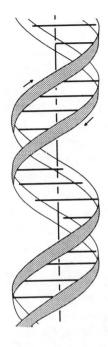

Watson and Crick's original schematic diagram of the double helix as it appeared in *Nature* in 1953. The horizontal rods represent the pairs of bases— that is, AT, TA, CG, or GC—which hold the chains together. (Reprinted with permission from *Nature*, Vol. 171, pp. 737–738. Copyright © 1953 by Macmillan Magazines Ltd.)

ray crystallography was created by Sir Lawrence Bragg in England. Bragg's objective was to use crystallography to determine the structure of giant molecules.)

Watson's quest to determine the structure of DNA eventually brought him to Cambridge University where he met Francis Crick. Crick had almost completed his doctorate in physics when World War II broke out. After seven years in defense-related research at the Admiralty, Crick decided to master biology. Judson has cited Crick's goals as noted in an application for a student grant in 1947:

> The particular field which excites my interest is the division between the living and the non-living, as typified by, say, proteins, viruses, bacteria and the structure of chromosomes. The eventual goal, which is somewhat remote, is the description of these activities in terms of their structures, i.e. the spatial distribution of their constituent atoms, in so far as this may prove possible. This might be called the chemical physics of biology."[16]

Upper left: The nucleic acid bases. The purines adenine (A) and guanine (G) are shown above with the pyrimidines cytosine (C), uracil (U), and thymine (T) below. *Lower left:* Hydrogen bonding produces the adenine-thymine and guanine-cytosine base pairings. (Reproduced with permission from *Recombinant DNA* by J. D. Watson, J. Tooze, and D. T. Kurtz, 1983. Scientific American Books.)

Even before Crick and Watson met, Crick had pondered how it would be possible for any genetic substance both to direct the synthesis of enzymes and structural proteins and also to specify how faithful copies of its information content would be passed along within each cell division during development and from generation to generation. Crick brought to the collaboration a stunningly brilliant and insightful mind, a dazzling understanding of the principles of X-ray crystallography, and an appreciation of both Pauling's work on the nature of the chemical band and Pauling's use of scaled three-dimensional representations of molecular components as "building blocks" to model the structure of complex molecules.

In a 1953 issue of *Nature*, James Watson and Francis Crick proposed the correct molecular structure for DNA with their formulation of the "double helix" hypothesis. This hypothesis was based on a series of ingenious insights by both men and their appreciation of the significance of the technically remarkable, high-quality crystallographic data on the structure of DNA obtained by Rosalind Franklin, an exceptionally gifted young woman working at King's College in London, England. From Franklin's data, Crick realized that the "backbone" of the DNA molecule consisted of two spiral phosphate-sugar chains coiling around one another and running in opposite directions. Watson discovered the rules for base pairing: adenine paired with thymine, guanine paired with cytosine. "The pairing could not be switched, however, for then the various atoms around the fringes got in each other's way. But when an A-T pair was laid on top of a G-C pair, the compound shapes were exactly congruent. Such pairs could fit inside the backbones without bulges or pinches."[17] The model immediately explained the significance of Erwin Chargaff's rules that the molar ratios of A:T and G:C were always one! In 1950 Chargaff had discovered that in all DNA thus far examined "the molar (that is, molecule-to-molecule) ratios of total purines to total pyrimidines, and also of adenine to thymine and of quanine to cytosine, were not far from 1."[18] The profound significance of his result that there was a molecule of adenine present in DNA for every molecule of thymine and similarly for the quanine-cytosine pair could now be appreciated. The genetic code itself was expressed in the sequences spelled out by the four base "letters" A, T, G, C.

Crick explained the discovery in a letter to his young son, Michael, away at school: "We have built a model for the structure of des-oxy-ribose-nucleic-acid (read it carefully) called D.N.A. for short. Our structure is very beautiful. It is like a code. If you are given one set of letters you can write down the others. Now we believe that D.N.A. is a code. That is, the order of the bases (the letters) makes one gene different from another gene (just as one page of print is different from another)."[19]

As for how DNA might replicate itself, Watson and Crick confirmed and extended Pauling and Delbrück's remarkable insight and ended their two-page paper in *Nature* with a classic understatement: "It has not escaped our notice that the specific pairing we have postulated immediately suggests a possible copying mechanism for the genetic material."[20]

With the publication of Watson and Crick's two-page note, the course of scientific history in our century was forever changed. In the first half of the century, the science of the physical world had been dominated by the ideas of Max Planck and Albert Einstein, Niels Bohr and Lord Rutherford, and their intellectual descendants. As the second half of the century opened, the biological world, stimulated by the ideas of Pauling and Delbrück and the model of Watson and Crick for the structure of DNA, would emerge as an equal partner.

Alzheimer's Disease at Midcentury

Even as molecular biology began its ascent toward its first golden age, neurology still struggled to define the relationship of Alzheimer's disease to senile dementia and of both syndromes to what was then called "normal senility." These were not trivial distinctions because for many years the incentive for scientists to study the dementias of later life and the financial resources available to them depended largely on whether they and society regarded these dementias as "diseases" or as inevitable consequences of the aging process.

Writing on this subject in the late 1940s, R. D. Newton, working at the Napsbury Hospital in London, began his influential paper by noting that in the four decades since the initial discovery of Alzheimer's "there has been little advance towards a solution of the problem."[1] I can find no evidence that Newton ever wrote again on the subject of Alzheimer's disease, but his exceptionally high standards and keen insights made his lone paper a classic. Newton discerned the outstanding issues regarding the causes of Alzheimer's disease with such clarity that his work served as a focal point for decades to come. As he reviewed the work of 40 years, Newton concurred with the majority opinion of the day that there were "no clinical, pathological or aetiological grounds" for distinguishing between Alzheimer's disease and senile dementia, other than noting the difference in age of onset. And even distinctions based on onset were often blurred or misleading. Some neurologists proposed age 55 while others suggested age 65 or even 70 as the cutoff point, for applying the term "Alzheimer's disease."

Newton accepted the view of his own teacher, the renowned British neuropathologist W. M. McMenemey, that the age limit "must be judged on its own merits in view of the known differences in aging in families and individuals." Newton cited examples where the diagnosis of senile dementia was made in a father followed some years later by a dementia with similar symptoms in his son. The son's dementia was diagnosed as Alzheimer's disease simply because his symptoms began a few years earlier than his father's.

Newton also summarized the previous reports of familial clusterings of Alzheimer's disease. The first reports had appeared in the German literature in 1932 independently by J. Shottky and by A. Von Braunmuhl, followed by a few studies in the British and American literature by the end of the decade. More reports followed in the 1940s. Taken together, these studies convinced Newton that an hereditary factor was preeminent in the development of both Alzheimer's disease and senile dementia. (There were then as yet no reports in the literature about Hannah's family.)

Newton further speculated that the primary cause of these dementias was the death of brain cells with the secondary formation of plaques (an essential, though not a specific, feature in Alzheimer's disease) and neurofibrillary tangles (an inconstant factor in Alzheimer's which was also found in various other brain diseases). If brain cells die, they do so, thought Newton, "because they have reached their allotted life-span. If this is so, then the whole question of an aetiology of Alzheimer's disease and senile dementia is probably a genetic one. The problem, therefore, becomes one of genetics."

Yet Newton stopped short of calling the genetic problem a disease. He was strongly influenced by the concept of a normal senility, an idea pervasive at the time. Newton quoted from an article written by Frederick Tilney of Columbia University in 1928: "Thus it appears, from the clinical as well as from the anatomical stand-

point, that senile dementia does not differ essentially from the ultimate state of normal old age except in its less [*sic*] rapid development."

Newton, then, believed that a severe and universal deterioration of mental processes inevitably occurred in advanced old age.
He also saw a continuum between presenile dementia, senile
dementia, and normal senility that led him to reconsider
Alzheimer's disease and senile dementia as "conditions" rather
than as "diseases." Thus, he proposed dropping the terms
"Alzheimer's disease," "senile dementia," and "senility" and
grouping all three together under the noncommittal descriptive
term "Alzheimer's dementia."

Newton's conclusion, whether correct or not, had focused on
the similarities between the symptoms and pathologies found at
three distinct life stages. Yet neither he nor any one else at the
time, apart from a recognition of obscure genetic factors, emphasized the importance of differences in etiology that caused the
Alzheimer process to take hold at these vastly distinct stages of
life.

Even so, Newton himself discovered a key biological difference
in the rate of development of senile plaques in males and females.
This rate may perhaps be a rough marker of the progression of
the process which over many years, even decades, leads to the
symptoms of dementia. In the United States, as early as 1934,
D. Rothschild had found that the incidence of Alzheimer's disease
was one and one-half times greater in females than in males. In
England a decade later this remarkable difference would inspire
W. Mayer-Gross's challenge: "A completely unexplained feature
of Alzheimer's disease is [its] prevalence [in] females."[2]

After his extensive review of the literature, Newton presented
his own results from his examination of the brains from 150 consecutive autopsies carried out on patients with psychiatric disorders who died in one or another of three mental hospitals near
London. Newton failed to find any senile plaques in the brains of
either males or females who had died before age 50. There were no
patients with presenile dementia in this age group in his study. Yet
when he examined the brains of patients dying between ages 50 and
64, he found some plaques in 11 percent of the males, but in minimal numbers, and in 29 percent of the females. Twenty-five percent of the affected females showed substantial changes. These
observations led Newton to propose that the onset of the
Alzheimer process, and the subsequent dementia, is scaled accord-

ing to the potentially longer reproductive life in the male. Curiously, these observations remain unexplained and unexploited even to the present day.

Yet other discoveries made during this period would be exploited later. In a brilliant series of studies from 1927 to 1934, Paul Divry in Belgium identified the "peculiar substance" in the senile plaques as "amyloid" (a then obscure class of substances defined by characteristic staining reactions).

Other discoveries were seen as significant only in retrospect. From a contemporary view, it is often very difficult for scientists to distinguish decisive clues from false leads. For example, in 1929 Friedrich Struwe discovered numerous senile plaques in the brain of a relatively young patient dying of Down's syndrome, or "Mongolian idiocy," as it was then called.[3] Then, in 1948, G. A. Jervis realized that not only did patients with Down's syndrome develop plaques, but also many of them developed a progressive dementia resembling that of Alzheimer's disease by their late thirties or forties.[4] Years later this premature expression of an Alzheimer-like process in Down's syndrome would demand an explanation.

As I look back at the period from Alois Alzheimer's initial discovery in 1906 to midcentury, I realize that it was not for want of effort, intellect, or imagination that progress in Alzheimer's disease was so slow. It was for want of effective techniques to apply to the problems. For over 50 years, neurologists had relied on their clinical examinations and neuropathologists on their observation of tissue under the light microscope. Neuroscientists had no technical advance comparable to X-ray crystallography which had opened the way for the solutions of protein structure and the structure of DNA. As the second half of the century began, many scientists from various disciplines studying Alzheimer's disease expectantly awaited comparable technical advances in their own field.

Enter Genetics

Within a dozen years after the publication of Newton's 1948 paper, which pointed to genetic factors in Alzheimer's disease, two notable attempts were launched to determine whether familial Alzheimer's disease was "linked" to then known genetic traits, or "markers." The first study focused on an afflicted family in England and the second on the descendants of Hannah and Shlomo.

This concept of linkage was a remarkable breakthrough marking the union of ideas from immunology, genetics, and statistics. Essentially, it stated that co-inherited observable traits, passed down from generation to generation, more often than can be explained by chance are determined by genes that are located close to each other on a common chromosome. This notion of linkage followed on the earlier discoveries of genetic linkage by William Bateson, the prolific and frequently controversial English proponent of Mendelism who gave the name "genetics" to his science in 1906, as well as by Morgan and Sturtevant. It had now evolved into a formidable instrument for gene mapping.

The basic contribution of immunology to linkage analysis can be traced back to 1875 when Leonard Landois in Germany observed that red blood cells from an animal of one species usually "agglutinated," or clumped together, when they were mixed with blood serum from an animal of another species. Immunologists recognized that this kind of clumping was analogous to the clumping that took place when bacteria were mixed with an appropriate immune serum.[1] In both cases the agglutination occurred when "antigens" on the surface of the cells united with antibodies in the immune sera. (An antigen is a substance that provokes the production of an antibody. Most human antigens are specialized subgroups of proteins or polypeptides that consist of highly specific sequences of amino acids.) In Landois's time, no framework had yet been established to relate these immunological results to the fundamental principles of genetics, which of course, except for Mendel's work, remained to be discovered.

It was not until 1900 that a 32-year-old physician named Karl Landsteiner, working in Vienna, observed the agglutination of human red cells when he mixed them with serum from individuals of the *same* species. He discovered that sera taken from some of his colleagues agglutinated the red cells of others.[2] Antibodies in some sera clumped red cells with an "A" antigen, whereas other sera clumped "B" antigens. Still other red cells were not agglutinated by either anti-A or anti-B antibodies; these red cells belonged to the "O" group that lacked both the A and B antigens. Thus, human red blood cells from different individuals could be typed as A, B, AB, or O according to the presence or absence of the A and B antigens. Landsteiner's discovery of the ABO blood group led to the introduction of blood typing and opened an era of safer blood transfusions which would save countless lives. Thirty years later, Landsteiner would receive the Nobel Prize for his discovery of the ABO blood groups.

Yet few in 1900 could have foreseen that Landsteiner's work would soon be recognized as having as profound a significance for genetics as for immunology. The next breakthrough would have to wait until 1908, when R. Ottenberg and A. A. Epstein would suggest that the A and B antigens might represent inherited traits. This was a brilliant inference, coming at a time when the renowned geneticist Bateson could still write in 1909 that, apart from eye color, Mendelian inheritance had as yet yielded scant evidence of any *normal* characteristics in humans.[3] Within a year, E. von

Dungern and L. Hirszfeld confirmed that Landsteiner's ABO blood types were inherited in accordance with Mendel's rules.[4]

But what if the ABO blood subtypes could serve as a set of markers for the site of a gene on a particular chromosome? It wasn't until later in the century that it would be recognized that the A or B antigens on the surface of the red cell, or the absence of them, were alternative forms of gene expression, or "alleles," at a single chromosomal site. Hence, when Landsteiner was detecting the presence or absence of a given antigen by the addition of the appropriate immune sera, he was in effect determining whether or not a person had inherited one or another allele, or subtype, of a given gene. Thus, the work of the early immunologists could now be tied directly to that of their colleagues in genetics. Sturtevant's construction of a crude "linkage map" in 1911 for genes on common chromosomes in fruit flies gave gene mappers the methods to estimate the odds or probability that any given trait was linked to any other trait, whether it be a physically observable characteristic such as eye color or a biochemical marker such as a blood group antigen.

All along, geneticists knew that specific diseases were linked to the X chromosome and were transmitted by the female parent to one-half of her male progeny. A female who carries an X-linked genetic defect remains clinically well because she is generally protected by a second and normal X chromosome that she has received from her normal parent. Hemophilia and color blindness result from such X-linked defects. How close together on the X chromosome were the genes for these two diseases? Now it was the turn of the statisticians to codify the rules for determining the relative proximity of the genes for linked traits. By 1939 in England, J.B.S. Haldane and J. Bell had inferred that these two defective genes on the X chromosome were closely linked. Although the early work focused on the X chromosome, the general statistical methods were applicable to the analysis of linkage on any chromosome.

The general principles remained straightforward when the co-inheritance of traits from a given parent to an offspring was frequent—the odds are greater that their genes lie close together on a common chromosome. Conversely, when the co-inheritance falls close to 50 percent—that is, when the two traits appear to be inherited independently or at random—then their genetic locations are not nearby and not linked. Such an absence of linkage would indicate that the genes either reside on different chromosomes or on widely separated locations on the same chromosome. By 1947 Haldane and C.A.B. Smith had worked out a formal mathematical model to estimate the odds within a given pedigree that two traits—or by direct implication their respective genes—were linked as a function of likely "genetic distances" between the locations of the two genes.

For a time gene mappers cited their odds in favor of linkage between two traits or genes in terms of ratios such as 100:1 or 1,000:1. Then, in 1949, G. A. Barnard introduced the term "lod" for logarithm of odds. This term is now universally used to characterize the statistical likelihood of linkage. A "lod score" of 3 indicates that the odds in favor of linkage are 10^3, or 1,000, to 1. Conversely, the odds that the apparent linkage occurred by chance are only 1 in 1,000. Hence, a lod score of 3 or greater is taken as strong evidence for linkage. By the end of the 1950s, C.A.B. Smith, S. M. Smith, and L. S. Penrose in England and Newton Morton in the United States had established the statistical basis for genetic linkage studies.[5] (Some specialists in genetics now recommend consideration of a "correction factor" for interpreting the statistical significance of lod scores, especially when evidence in favor of linkage is marginal.[6])

Even up to 1950, linkages had been established for genes on only two of the forty-six human chromosomes, namely, on either the X or Y sex-determining chromosomes. As yet, no single example of linkage of traits in humans to any "autosome" had been established. That is, no link had been made to one of the 22 pairs of chromosomes not involved in determinating sex. (The autosomes are numbered according to chromosome length; thus, the longest is chromosome 1. However, the shortest autosome is chromosome 21, not 22. These two shortest autosomes were initially misnumbered, but later, when the error was discovered, geneticists deemed it too late to modify an accepted convention.)

The first linkage to an autosome was not made until 1951. In that year J. Mohr in Norway linked an antigen (known as the Lutheran factor after the patient who donated the blood) to the protein product of another gene. Even though the parent chromosome of the two linked genes had not yet been identified, Mohr's finding encouraged geneticists that discoveries of more examples of autosomal linkages would follow and that the genetic linkage maps of the autosomes would eventually become possible.

The tacit assumption that a trait or genetic defect would always be linked to only one gene was shaken in 1953 when Sylvia Lawler and her colleagues demonstrated linkage between the genes for the Rh blood group factor and the gene responsible for oval-shaped red cells—elliptocytosis—in some families. Here was an example of genetic heterogeneity. That is, a different gene—in this case one not linked to the Rh gene—had to be responsible for the oval shape in other families. The seemingly mundane observation would have immense importance for future gene mappers.

Now, in the early 1950s, over half a century had passed since Landsteiner discovered the ABO blood groups and almost 40 years since geneticists had recognized that the blood group antigens could be used as genetic markers. Even so, no traits or genetic defects had yet been linked to these markers. Then in 1955, J. H. Renwick and Sylvia Lawler, working in London, established the first linkage of an autosomal syndrome, the "nail-patella syndrome"—a rare disorder characterized by an abnormal development of the nails and an absent or underdeveloped patella—to the ABO blood group. It would still be decades before the genes for the ABO group were localized to a specific chromosomal site—the long arm of chromosome 9.[7]

It was against this background that Lorna Wheelan, also working in London in the late 1950s, made the first attempt to determine whether familial Alzheimer's disease could be linked to any of the then known blood groups. Her study was also the first to search for linkage of a genetic brain disorder that did not manifest itself until middle life. Wheelan chose to study a family from a London suburb in which a mother and five of her ten children had developed Alzheimer's disease in their middle to late forties, at ages of onset similar to those in Hannah's family.

All descendants of the deceased mother were traced, but only two affected offspring were still living. Robert R. Race, a leading expert on the inheritance of blood group antigens, was called in to examine the blood samples from all the living descendants of the mother. Based on the known blood types of her children and other relatives, Race was able to "reconstruct" each individual's "genotype" with respect to his or her blood group; in other words, he could infer the identity of each person's two gene copies. Race then tried to determine whether the mother's gene for Alzheimer's disease was passed on to her affected descendants, along with one or another of her red blood cell antigens. To do this, he would need to test each of the blood group antigens for linkage to Alzheimer's disease as that defective gene was passed from the mother to her similarly affected children. Race concluded that the genes for the ABOs and Rh blood types were inherited independently of the Alzheimer gene.

But, in the midst of his work, a curious and previously discovered minor blood group emerged as a possible nearby marker for a gene causing familial Alzheimer's disease. This blood group, the MN system, consisted of two inheritable antigens M and N, one of which differs from the other as a result of some remote mutation. While these antigens did not have to be matched when blood was typed for safe transfusion, their antigenic differences could be useful in detecting nonpaternity—that is, in excluding a given man as the biological father of a specific child. Moreover, these antigens, just as those of the ABO and Rh groups, were already in use in anthropological research which traced ancient migrations and the roots of the various modern-day branches of the human family. However, for Robert Race and Lorna Wheelan, the importance of the MN system lay in the fact that these antigens were markers for nearby genes on some then unknown chromosome.[8] Two other blood group antigens, designated as S and s, were controlled by alternate expressions of a nearby gene. Together these four antigens were designated as the MNSs system.

When Race studied the MNSs system, he found that the mother had given her gene for one red-cell antigen and her gene for Alzheimer's disease to two children and her non-Alzheimer gene and her genes for a different antigen to two other children who were now well beyond the age of risk and were thus unaffected. Race could only conclude that the results were due to genuine linkage or to chance. Since the number of individuals in his study was so small, the results could easily have been by chance. As he emphasized, "It must be made clear that this apparent linkage may be due to chance. Only the study of other families suffering from the disease will decide whether the Alzheimer and the MNSs genes are in fact carried on the same pair of chromosomes."[9]

Wheelan and Race emphasized both the potential opportunities and the harsh limitations of linkage methods for locating the sites of human genetic diseases. In their work, the odds were slightly in favor of linkage, but not compellingly so. Geneticists searching for a mutation in a single human family were powerless to go beyond the data provided by the family under study; after all, they could not increase their sample size at will as Morgan and Sturtevant had done with their fruit flies. The geneticist studying human disease could only propose a marker as a tantalizing candidate for linkage to an important disease mutation. The hope was that some future gene mapper, perhaps one with access to a larger family or a more informative nearby genetic marker, could either confirm or disprove linkage to his previously proposed candidate site.

A second attempt to test possible linkage between familial Alzheimer's disease and blood group antigens was initiated by Hannah's grandson Charles on his own family in 1960. Charles, age 40, having earned a prestigious position as a pathologist at a distinguished Midwestern medical center, was approaching an age at which risk becomes only too apparent. The idea for the study came to him as soon as he read the paper on Alzheimer's disease by Wheelan and Race. Charles knew that his family had more living affected victims and more who had escaped the disease than was true of the English family. Thus, he had the opportunity to obtain statistically more robust data than had Wheelan and Race. Charles must have been tantalized by Race's suggestion that the disease might be linked to the MNSs system. Knowing, however, that he did not have access to all the antibodies that would be required if he were to carry out the study himself, he brought the project to the attention of Tibor Greenwalt, a noted expert on blood groups and an editor of the journal *Transfusion*. Greenwalt, with the help of his technician Thomas Sasaki, decided to perform a systematic test for linkage of Alzheimer's disease in Charles's family against all known blood groups.

Charles collected blood samples from 75 members of his own family and assigned a status to each—"affected," "escapee," "at risk," or "unaffected spouse." With this map in hand, Greenwalt could test for possible linkage. Greenwalt could carry out this phase of the study in two ways. First, he could set aside the results on Charles's own blood and then, only after he had determined whether or not there was linkage to a given blood antigen, he could look at the data on his co-investigator's blood group. Second, Greenwalt could include Charles's personal data with those on the rest of his family. The first choice would defer Charles's personal anxiety, revealing the implications for his own future only if Greenwalt established linkage. The second choice would have been more scientifically correct and more informative for reconstructing Hannah's blood type, but it could have represented a constant source of anxiety for Charles. With regard to this alternative, if Greenwalt had found tight linkage to a blood group marker, he would have known immediately whether or not Charles himself had a high probability of carrying the abnormal gene for Alzheimer's disease.

Was Charles aware of the implications for his own life as Greenwalt began analysis of the data? I believe that Charles had to have been aware, for he had read Wheelan's paper, and she had clearly stated, "Secondly, in a disease such as Alzheimer's where the effects are produced late in life after marriage and procreation, the possibility of knowing at an early stage, by means of the linkage, which family members will become affected would be of practical value, and advice about ordering their future life could be given."[10]

There are yet other reasons for believing that Charles must have known that he was about to play with fire for a second time. For Charles had already discovered that a familial hyperlipidemia predisposing to heart attacks by the early fifties or sixties also ran in his family. He had previously tested his family to determine whether the hyperlipidemia present in many of Hannah and Shlomo's descendants was associated with Alzheimer's disease. Charles found no association between the two diseases, but when he tested his own serum, he discovered sadly that he carried the lipid abnormality for which there was then no known effective treatment.

Wondering how Greenwalt had gone about the linkage study, I obtained and reviewed copies of his original records. I found that he had listed Charles's name and the blood-typing results directly in their proper places along with those of Charles's affected brothers and sisters. At each stage, as Greenwalt tested every successive blood group for linkage, he would know—if he found a linkage—whether or not Charles would likely be affected. Even then, geneticists correctly surmised that the antigens of these different blood groups were products of genes located on at least several different chromosomes. Thus, given the chromosomal locations of all the antigens tested, Greenwalt could test for linkage over almost 10 percent of the genome. The chances were small, but not remote, that a linkage could be found.

By early 1961 Greenwalt had tentatively concluded that Alzheimer's disease in Charles's family failed to link to any blood group. Still, Wheelan and Race's results, specifically on the MNSs system, must have continued to play on Charles's mind. In his notes of December 26, 1961, Greenwalt summarized the results of his retesting of the MNSs blood group on the key members of Charles's family. Greenwalt again failed to find any evidence linking Alzheimer's disease to the MNSs blood group. Charles, at last, could be satisfied that familial Alzheimer's disease, at least in his family, did not link to a known blood group. The period of daily wondering over whether Greenwalt would find linkage—and in the process perhaps predict Charles's own fate—had passed. With this acute stress over, Charles now had to endure a prolonged period during which he would not know whether he carried the gene for Alzheimer's disease until he either developed symptoms of the disease or lived past the age of risk. With the techniques then available, little more could be done to study the genetics of familial Alzheimer's disease.

Charles still thrived. With indomitable courage and firm determination, this man who could look at his own mortality and tempt fate without flinching was resolved to continue the fight.

New Seeds Are Sown

By 1960, following completion of his genetic study with Tibor Greenwalt, Charles knew that further genetic linkage studies on his family at this time were unlikely to be rewarding given the paucity of available genetic markers. He surmised that the best available research option might be the study of RNA or protein synthesis in Alzheimer's disease inasmuch as these biochemical processes were directly controlled by the unknown Alzheimer's gene. Following the literature on these topics intently, Charles came upon the work of Saul Korey, a neurologist and neurochemist working at the Albert Einstein Medical School in New York. Charles resolved to bring his family to Korey's attention and hoped that he could convince Korey to begin neurochemical studies of Alzheimer's disease.

Charles's timing could not have been better. Robert Terry, then a young morphologist and electron microscopist also at Einstein, told me that he and Korey had already decided in 1959 to target Alzheimer's as one of the brain diseases they would attack in a joint research program. Terry was eager to exploit the recent

advances in electron micrography which permitted scientists to
obtain images of biological structure through great magnification
and high resolution. If they were to distinguish disease-specific
abnormalities from the myriad of structural and chemical changes
that occur upon the death and dissolution of brain tissue, both
Terry and Korey would need to limit their approach to certain diseases. Specifically, they decided to zero in on those diseases that
could be studied by analysis of small samples of fresh brain tissue
that could be obtained at the time of brain biopsy according to
strict medical and ethical standards.[1] Hence, they selected two
classes of diseases to meet their criteria: the diffuse dementias such
as Alzheimer's, then considered a rare presenile disorder; and the
inherited lipid storage diseases such as Tay Sachs disease, in which
an enzyme deficiency results in the accumulation of a fat-like substance which in time destroys the nervous system.[2] With respect to
Alzheimer's disease, Terry hoped that examining the fine structure
of the plaques and tangles could provide further clues as to their
cause, or at least point his colleague toward the most promising
structural candidates for neurochemical studies.

Such was the plan when Charles first met Saul Korey, and
Korey was in agreement with Charles's proposal. With Robert Katzman, a young neurologist who was just beginning his research
work, and Kunikiko Suzuki, another young neurologist, Korey
began chemical studies of brain tissue from three patients with
Alzheimer's disease and compared these results with data from
unaffected controls. One of their three patients was Charles's sister, Sarah, who died of her disease while the study was under way.
Sarah was survived by a son and a daughter. (Her son was Jeff who
became my patient 25 years later.)

Korey's group first focused on several neurochemical issues.
Applying biochemical methods, they investigated the rates at
which specimens of tissue taken from the brains of affected
patients and controls consumed oxygen and energy stores. Here no
abnormalities were found. Then they studied the chemical components of cells in "gray matter" (the cerebral cortex itself composed
of nerve cells and neuroglial cells) and "white matter" (that great
mass of "axons" or nerve fibers connecting diverse brain regions).
Certain differences were found between the normal and abnormal
tissue. However, the researchers came to believe that most, if not
all, of these results were *secondary* to the known loss of brain cells
and subsequent changes during the course of Alzheimer's disease
and did not reveal its primary cause. These workers faced the stark

reality that even the most careful chemical study could not distinguish disease *cause* from disease *effect*.

By now, Terry was carrying out his electron micrographic studies on tissue from the same three patients that Korey's group had studied neurochemically. Terry published his work on the first two patients in 1963. The following year, Terry along with Nicholas Gonatas and Martin Weiss, published their comprehensive paper titled "Ultrastructural Studies in Alzheimer's Presenile Dementia," which documented the changes found in Sarah's brain.[3] The ultrastructural studies described the appearance of the amyloid in the plaques which appeared both in the spaces between nerve cells and within various nerve cells and their branches called dendrites. They also described the fine structure of the neurofibrillary tangles, which appeared to be composed of dense bundles of neurofilaments; they were normal in appearance apart from their vastly increased numbers. These bundles were packed so tightly within nerve cells that they seemed to "crowd out" the normal specialized structures within the cell. Terry's group also proposed that the neurofilaments in the tangles were not composed of the same substance as the amyloid in the plaques.

In London, at about the same time, Michael Kidd found that the appearance of the filaments depended on the thickness and "obliquity" or slant of the section examined with respect to the horizontal axis of the filament. Looking at near horizontal sections under very high magnification, he realized that the filaments were tightly wound double helices, "paired helical filaments" as he called them.

Korey's and Terry's work on the lipid storage diseases was progressing apace, when Korey died unexpectedly. But it was through this tragedy that Terry abandoned work on these diseases, which could not be pursued without the expert collaboration of a neurochemist,[4] and thus decided to concentrate on Alzheimer's.

With Korey's death, Katzman and Suzuki were left to complete the chemical studies of the Alzheimer brain alone and publish the work. Their paper, titled "Chemical studies on Alzheimer's disease," appeared in the *Journal of Neuropathological Experimental Neurology* in 1965.[5] Meanwhile, Katzman had turned his attention to a then recently described and frequently treatable dementia caused by "normal-pressure hydrocephalus." In this condition the ventricles, the cavities within the brain, enlarge as a consequence of excessive accumulation of cerebrospinal fluid and impair brain function.

In performing biopsies on these patients, Katzman found that many of the patients previously thought to be suffering from normal-pressure hydrocephalus were actually suffering from Alzheimer's disease.[6] These observations, together with his own personal experience—his mother-in-law suffered early-onset Alzheimer's—mobilized his energies to a lifetime interest in this disease and to providing social support for its victims and their families.

As the decade progressed, Charles continued to bring affected members of his family to Terry and Katzman's attention. Both researchers benefited from the contributions of Charles and his family to their research programs, and both recognized Charles's brilliance and his research contributions in his own field. But, as Terry told me, from that point on Charles chose to conduct his own subsequent basic research outside the field of Alzheimer's disease. "He felt too close. Besides, he knew he might soon develop the disease." By this time, the late 1960s, Charles had reached the mean age-at-risk for the disease as he had defined it a decade before. Laura, one of Charles's grand-nieces, remembered from family stories how difficult this period was for him. "Whenever he forgot something, he became terribly worried that he was developing Alzheimer's." Terry, too, remembered how concerned Charles became whenever he forgot where he had left his pen. Although Terry didn't worry about Charles's minor lapses, he was concerned about the 50-50 genetic risk Charles faced. Terry concealed his own apprehension, however, and sought to reassure Charles, "Stop worrying, I've misplaced my own pen three times today already."

Of course, everyone forgets now and then, even in youth, some more frequently than others, and there is a mild age-associated memory impairment not associated with Alzheimer's disease that affects virtually everyone by age fifty.[7] Yet in their earliest stages the "benign" minor lapses of normal aging cannot be distinguished from the earliest symptoms of Alzheimer's disease. Charles could not predict which would be his destiny, but, despite his fears, he remained determined to continue his own work and to help researchers on Alzheimer's disease whenever possible and for as long as possible.

When I think of Charles at the close of this stage of his life, I am reminded of Dr. Bernard Rieux in Camus's *The Plague*. The plague that beset Dr. Rieux's city came to an end and Rieux compiled a chronicle so that he "should bear witness in favor of those plague-stricken people; so that some memorial of the injustice and outrage done to them might endure and to state quite simply what we learn in a time of pestilence: that there are more things to admire in men than to despise."

The plague of familial Alzheimer's disease had not then and has not yet come to an end, and Charles wrote no chronicle of the events that I have thus far described. Yet had he done so, I believe that his closing words would have echoed those of Camus's Dr. Rieux. His chronicle, like Rieux's, "could be only the record of what had to be done and what assuredly would have to be done again in the never-ending fight against terror and its relentless onslaughts, despite their personal afflictions, by all who, while unable to be saints but refusing to bow down to pestilences, strive their utmost to be healers."

No Longer Alone

A disease such as Alzheimer's is an especially cruel affliction, devastating alike to its victims and their families. The families are rendered almost helpless to respond to the inexorable downhill course of their loved ones, who eventually lose their ability to preserve those meaningful human contacts that had bound their lives together. Ben had suffered these losses repeatedly. Now nearing age 60 as the 1960s drew to a close, he still carried the added burden of loneliness and isolation born of a belief that his family was one of only a few in the world afflicted with a familial form of Alzheimer's disease.

In the fall of 1969 after the Ciba Foundation held a symposium on Alzheimer's disease in London, Ben was astonished to learn how widespread such familial cases were. The conference organizer, G. E. Wolstenholme, had prophetically chosen a broad title for the meeting—"Alzheimer's Disease and Related Conditions"—in recognition of the uncertain boundaries of the disease.[1]

From a paper by R.T.C. Pratt, Ben learned that genetic forms of the disease were being studied in England, Germany, Sweden, Switzerland, and Belgium, as well as in the United States and Canada. As early as 1950, Franz Kallmann had realized the importance of genetic factors in dementias of later life by studying twins. He found that, if one identical or "monozygotic" twin (both twins deriving from one fertilized egg) develops dementia, then the chances that the other develops the condition are 43 percent; whereas if a fraternal or "dizygotic" twin (the twins being the products of two separately fertilized eggs) develops the disease, the chance that the other will develop the disease is only 8 percent. The much higher concordance of the disease in identical twins implied the primacy of a genetic component since in his study members of both types of twins shared common postnatal environments.

In Sweden large-scale studies on the general population, begun in the early 1950s by T. Sjogren, H. Sjogren, and A. Lindgren and continued through the early 1960s, suggested a genetic susceptibility factor even in late-onset dementia. From Switzerland, equally careful studies in the early 1960s by J. Constantinidis, G. Garrone, and J. de Ajuriaguerra were in general agreement with the Swedish work. Moreover, the Swiss workers found a number of cases in which one sibling developed the early-onset form of Alzheimer's disease and the other sibling developed the late-onset form. The clinical symptoms and the brain pathology in the two patients were essentially the same. Did it make any sense, then, to diagnose the first case as Alzheimer's disease and the second as senile dementia?

By the mid-1970s, as Robert Katzman thought about the significance of recent results, he proposed a new conceptual framework relating senile dementia to Alzheimer's disease and considering both as diseases distinct from normal aging. First, he realized that senility is not an inevitable concept of advanced old age, noting "that functional integrity in extreme old age is not confined to an Adenauer, a Picasso, or a Casals as has been shown by the Duke longitudinal study."[2] Second, although he agreed with R. D. Newton that there was no way to distinguish cases of presenile and senile dementia apart from the age of onset of the presenile before age 65, he boldly challenged Newton's concept that both were "conditions" that should be grouped together under the term "Alzheimer's dementia." On the contrary, he maintained both were "diseases" that were distinguishable from normal aging. With Toksoz Karasu, a psychiatrist interested in similar issues, Katzman wrote: "We should like to make the suggestion, simplistic as it may be, that we should drop the term 'senile dementia' and include these cases under the diagnosis of Alzheimer's disease."[3] Katzman then examined existing epidemiological data on the incidence of dementia in the aging population (which was growing greatly in number because of increasing life expectancies) and estimated the percentage of these cases due to Alzheimer's disease. He was astonished to realize that Alzheimer's disease had become the fourth or fifth most common cause of death in the United States.

Katzman brought his ideas together in a succinct two-page editorial that appeared in the *Archives of Neurology* in 1976.[4] His closing sentence was both an alarm and a call to arms:

> In focusing attention on the mortality associated with Alzheimer's disease, our goal is not to find a way to prolong the life of severely demented persons, but rather to call attention to our belief that senile as well as presenile forms of Alzheimer are a single disease, a disease whose etiology must be determined, whose course must be aborted, and ultimately a disease to be prevented.

Ben had called that Ciba Symposium of 1969 "our first inkling of real hope. Perhaps if enough people were interested in the disease we could find a cure." With Katzman's editorial, that interest would come.

Twin Pillars of Hope

Katzman now emerged as a principal advocate for more funding for both research on Alzheimer's disease and social support for its victims and their families. In the mid-1970s he launched a two-pronged attack, encouraging more interest at the governmental level while working to develop organizations that would promote such support at the community level. His editorial suggesting that both the presenile and senile forms of dementia be called "Alzheimer's disease" could not have come at a more propitious time.

Strong support for medical research from the federal government was then relatively recent. In fact, according to Patrick Fox, only after the successful development of the Salk and Sabin vaccine for polio in the 1950s did the public's faith in the practical value of medical research become firm. By the late 1940s Mary Lasker and Florence Mahoney had developed an effective strategy for raising money for biomedical research based on targeting specific dreaded diseases. Fox credits these two women as being "largely responsible for the emergence of the NIH from a relatively obscure division of the Public Health Service, primarily involved in cancer research, to the world's largest medical research center."[1] Lasker, who with her husband founded the Albert and Mary Lasker Foundation to support biomedical research, emerged as a leader in the fight against cancer, and Mahoney as an advocate for research on mental health and aging.

A reporter and the wife of the publisher of the *Miami Daily News*, Florence Mahoney originally came to Washington to cover events in the Capitol for her husband's paper.[2] After his death, she inherited his fortune and resolved to use her financial resources and energies to advance causes to which she was dedicated. A woman of many interests, she significantly helped advance the development of the National Institutes of Health and for a time served on the National Institute of Child Health and Human Development Council. She also became active in the Democratic party, supporting in turn Hubert Humphrey in the primaries and John Kennedy in the general election in 1960. By the early 1970s Mahoney strongly advocated the creation of a national institute on aging.

The legislation that created the National Institute of Aging (NIA) was passed in 1974, and Mahoney recommended that Robert Butler, a psychiatrist and the author of the newly published book, *Why Survive? Being Old in America*, be appointed director of the new institute. On the day that he became director, Butler was awarded the Pulitzer Prize for his influential book. Despite this auspicious start, Butler knew that his work would not be easy; and later, he recounted to me his troubled early days as director. He recalled that many medical scientists had been hostile to the founding of his new institute, convinced it might drain away desperately needed research dollars from established and perhaps far more attainable goals. Most regarded attempts to counter the aging process as attempts to curtail the inevitabilities of life rather than as efforts to understand specific diseases.

Butler spiritedly devised his counterattack. He unabashedly identified his key strategy as the need to practice "the politics of anguish." He understood the problem only too well: "Congress doesn't vote funds for basic research; they vote to fund research on specific diseases." Thus, Butler envisioned the need for a categorical disease to be used as the wedge for obtaining increased funding from Congress for his new institute. Katzman's more inclusive definition of Alzheimer's disease would serve as that wedge.

Concurrently, Katzman tried to organize a lay organization in New York City dedicated to Alzheimer's disease. Katzman asked

Leonard Wollin, a New York attorney and a member of a family afflicted by Alzheimer's disease, to help him found such an organization. Only after the third member of Wollin's family had died of Alzheimer's did he become convinced to go ahead with Katzman's plan.[3] The Alzheimer's Disease Society was incorporated in December 1978.

Meanwhile, Katzman and Terry had begun corresponding with Donald Tower, then director of the National Institute of Neurological and Communicable Disorders and Stroke. In April 1976, according to Fox, they sent a copy of Katzman's editorial to Tower, with a letter suggesting that a conference be held to discuss the problems of dementia and to suggest the development of a research center on Alzheimer's and related diseases. The conference was held in June 1976. Following the meeting, Butler urged the formation of a voluntary health organization devoted to Alzheimer's disease and related disorders. Butler and Katzman would soon learn that a number of important regional organizations dedicated to helping such patients and their families were springing up across the country.

In 1977 Roberta "Bobbie" Glaze Custer of Minneapolis had become a widow at age 59 after almost a decade of caring for and grieving over her husband Ken diagnosed as having Alzheimer's disease.[4] Though Ken's illness had been devastating both financially and emotionally, Custer decided that she had to find a way to fight back and help others similarly confronted. The full impact of Katzman's editorial identifying senile dementia as Alzheimer's disease had not yet been felt, and Custer could find few in her community who had even heard of Alzheimer's. For a time she searched for Alzheimer families, with no success. But then a reporter for the *Minneapolis Star and Tribune* contacted her for an interview. To find others like herself, Custer knew that she had to tell her story with complete candor. The paper gave her interview a full page, using an apt and unforgettable quote as its headline: "The Funeral That Never Ends."

When the story broke, five afflicted families in Minneapolis contacted her, and after the Associated Press picked up the story, Custer began receiving letters from around the country. She and her two newly found Minnesota allies, Hilda Pridgeon and Muriel Erickson, dug into their pockets to put down the $20 needed to open a checking account for their new organization. Custer vividly recalled, "It began at my kitchen table," as she described the events that led to her contacts with others who would now start their own support groups around the country. In November 1978 she traveled to Washington to receive the National Center for Voluntary Action Award for her program called REACH, an acronym for Reassurance to Each, from the National Mental Health Associations.

In September 1979, Custer received an invitation to attend a symposium on dementia sponsored by the University of Minnesota and the Veterans Administration. Contacting caregivers from lay organizations in Seattle, San Francisco, New York City, and Columbus, she asked if they would join her at the meeting and discuss the possibility of organizing a national association.[5] Each group sent a representative. Moreover, Dr. Butler sent his assistant Marian Emr, who reported back to NIH on the enthusiasm of the lay workers.

Butler invited Bobbie Custer and other lay leaders to NIH the very next month to meet with him and Katzman. Custer recounted to me the names of the chapter leaders—Martha Fenchak (now Martha Bell) from Pittsburgh, Anne Bashkiroff from San Francisco, Warren Easterly from Seattle, Dr. Leopold Liss from Ohio State, Lonnie Wollin from New York, Dr. Marrott Sinex from Boston University Medical Center, and Hilda Pridgeon and Bobby from Minneapolis. Also invited but initially declining the invitation was a Chicago businessman named Jerome Stone.

In 1970 at age 50 Jerry Stone's wife was diagnosed at the University of Chicago as suffering from some unspecified "presenile dementia." Stone took his wife to Massachusetts General Hospital in Boston where the eminent neurologist Raymond D. Adams told Stone that his wife suffered from a rare disease—Alzheimer's. Stone told me that he went to a medical library, took out ten books on neurology, and searched for information on Alzheimer's disease. The longest description he would find was one and one-half pages. Between 1973 and 1974 Stone searched for opportunities to promote research. He eventually found his way to Robert Terry at Einstein and offered his help. Through Terry, Stone met Katzman and Wollin. Stone was the chief executive officer of a major container company, and Terry and Katzman recognized how important a man of his organizational talents would be for a planned national organization. Thus it was that Stone was invited to the October 1979 meeting at NIH. Since Stone was on his way to China for a combined business-pleasure trip, he declined but he did agree to attend a dinner that Florence Mahoney had arranged the night before the NIH meeting. At this dinner Stone met the other invitees, recognized that the correct mix of people and ideas had come together at

Voices of advocacy. *Upper left:* Dr. Robert Katzman. (Courtesy of Robert Katzman.) *Upper right:* Dr. Robert Butler. (Courtesy of Robert Butler.) *Lower left:* Jerome Stone. (Courtesy of Mr. Stone.) *Lower right:* Dr. David A. Drachman. (Courtesy of Dr. Drachman.)

a crucial time, realized that his organizational skills and business acumen would be needed, and canceled his trip to China.

At this meeting the future organization chose its name—the Alzheimer's Disease and Related Disorders Association, or ADRDA. Losing no time, Stone agreed to host its first official meeting two months later in Chicago in December 1979. At this meeting Stone, whom Bobby Glaze Custer calls "the wisest choice we could have made," was elected president. Katzman agreed to serve as chairman of the Medical and Scientific Advisory Board. My colleague, Dr. David Drachman, would succeed him in this position in 1985. Custer accepted appointment as National Program and Chapter Development chairman. Stone recalled to me his new organization's objectives subsumed under the acronym "RECAP" to include research, education, chapter development, and patient care.

In the coming years, the various regional groups fought over the scope and character of the organization. Katzman was fearful that too wide a scope would dissipate potential resources; however, he remained sympathetic to the broader concerns. Fox writes:

> Katzman's viewpoint was that the services of ADRDA-sponsored family support groups should be available to anyone, regardless of disease. Public education and scientific research, however, were a different matter. His view was that these activities should focus only on Alzheimer's disease and a limited number of researchable related disorders which might illuminate the characteristic of the disease."[6]

Eventually, Katzman's view that research be targeted toward Alzheimer's disease prevailed.

Meanwhile, Butler hired Zaven Khachaturian, a neurobiologist with a strong interest in brain chemistry, to establish a neurobiology of aging program. His job, Khatchaturian told Fox, "was to work with the scientific community and find good scientists who could be turned on by the neurobiology of aging and Alzheimer's disease." Butler continued to focus his own efforts "on representing NIA to Congress and in developing a public constituency to be an advocate for the Alzheimer's disease research cause."

Within six months, members of the new group were ready to testify before Senator Thomas Eagleton's Committee on Labor and Human Resources on behalf of the thousands of families that had now been identified. Bobbie Custer was selected to testify on the plight of the disease victims and their caregivers. She recalled the response of the very deeply moved and sympathetic Eagleton, "Bobby, it looks like we have a hell of a job ahead of us."

Public awareness of Alzheimer's was given another boost in October 1980 when Abigail Van Buren printed a letter in her nationally syndicated column "Dear Abby" about a family's severe difficulties in managing an Alzheimer's victim. In her reply Van Buren referred her correspondent to the ADRDA for assistance and information. As a result of all the ensuing media attention, the newly emerging organization gained wide public visibility.[7] In fact, the organization received between 30,000 and 40,000 letters. With this nationwide publicity, Alzheimer's had at last become a household word and ADRDA had secured a sure footing. Mutually supportive, the NIA and the ADRDA could now reach toward their objectives.

Though not among the founders of ADRDA, several of Hannah's distinguished descendants served on the various advisory boards of the new organization. Their main interest, as it had been in the past, remained in working directly with the scientists who were at the forefront of research. And even as ADRDA was being formed, new approaches to the causes of Alzheimer's disease were being actively pursued.

Use of the unified term "Alzheimer's disease" to encompass both presenile and senile forms of the disease had galvanized public awareness and had led to the founding of ADRDA as well as to the support of the NIA. But Katzman, Butler, and many scientists continued to entertain doubts as to whether the causes of the early and late forms of the disease were really identical. Several commissions were set up in the late 1970s to propose promising research areas and to recommend an unbiased terminology for the early and late forms of the disease. So as "not to prejudge the etiological identity of the disease at different ages," the commission that included Robert Katzman and Robert Butler recommended that the disorder retain the term "Alzheimer's disease" to describe the presenile form but "senile dementia of the Alzheimer's type" for the senile form. For several years after the publication of this report,[8] neurologists and psychiatrists frequently employed the term "senile dementia of the Alzheimer's type." However, in time, more as a matter of convenience, the more precise term for the late-onset form came to be abbreviated as simply "Alzheimer's disease." In hindsight, it appears that Katzman's brief article in 1976 recommending the term "Alzheimer's disease" for both presenile and senile forms of the disease remained more influential than his commission's more carefully considered recommendations in 1978.

A Transmissible Virus?

By the mid-1960s Carlton Gajdusek had ample reason to wonder whether some forms of Alzheimer's disease could be caused by an infectious agent. More than once in his career, he had discovered that certain brain diseases then thought to be genetic were in fact caused by infection. A decade earlier, while still in his early thirties, this courageous and remarkable young man was investigating the newly discovered and mysterious, invariably fatal brain disease called "Kuru." This disease was the scourge of the Fore people who still lived in a Stone Age culture in the Eastern Highlands of New Guinea. Taking personal risks that would have astonished even a young Arrowsmith, Gajdusek lived side by side with the native population sharing their exposure to infectious diseases, perhaps even to Kuru, and forging personal bonds that would last a lifetime.

Gajdusek was well prepared for the incredible detective work that lay ahead. As a high school student, he acquired a sound background in mathematics, chemistry, and physics as well as in the biological sciences. After attending the University of Rochester, Gajdusek entered Harvard Medical School while still a teenager. Later, he received clinical training in pediatrics and studied virology under John Enders, the Nobel Laureate whose early work on tissue cultures had opened the way for the successful polio vaccines. When Gajdusek visited New Guinea in early 1957 and first observed the victims of Kuru, he was a fellow of the National Foundation for Infantile Paralysis working under subsequent Nobel Laureate Sir MacFarlane Burnet, the renowned immunologist in Melbourne, Australia. The New Guinea territory was administered as a trust by the Australian government, and the medical officer for the region, Dr. V. Zigas, had first learned about the disease in 1955, following early reports from patrol officers and anthropologists.

Kuru was also the Fore word for "the trembling associated with fear or cold." "The shaking disease," as government officers called it, was the major medical problem in the area. The disease begins with the gradual onset of ataxia, uncoordinated limb and body movement, followed soon after by an involuntary tremor. Within months, patients can no longer walk or stand without substantial support. Soon thereafter, victims are unable to maintain their equilibrium even in a sitting position. In the early months of the illness, speech becomes slurred and finally unintelligible, with a progressive decline of intellectual function leading to total incapacitation. Kuru was often fatal within three to six months of onset, with victims rarely living beyond a year. Initially, Gajdusek and Zigas could find no evidence of an infectious or postinfectious disease. Studies of blood and spinal fluid showed none of the tell-tale immunological changes that suggest the presence of conventional infectious agents. Thus, at first, Gajdusek and Zigas, along with Australian public health officials, believed that Kuru might be a genetic disease or at least that there might be some hereditary predisposition for the disease.

The Fore people believed that Kuru was the result of magic or sorcery worked upon the victim by an enemy. At the time Gajdusek arrived, the tribes had been pacified for only the past five years, and the government had taken measures to end hostilities between tribes and villages and to discourage sorcery and cannibalism. Even so, sporadic warfare and cannibalism persisted, and ritual killings were carried out as reprisal for Kuru magic.[1]

In time, Gajdusek came to suspect that a novel transmissible agent—spread at the time of ritual cannibalism, when survivors were bound by tradition to prepare and then ingest brain tissue from their deceased relatives—was the cause of Kuru. As Gajdusek later could happily note in his Nobel address, published in 1977:

> The incidence of the disease in children has decreased during the past decade, and the disease is no longer seen in either children or adolescents. This change in occurrence of kuru appears to result from the cessation of the practice of ritual cannibalism as a rite of mourning and respect for dead kinsman, with its resulting conjunctival, nasal and skin contamination with highly infectious brain tissue mostly among women and small children.[2] [Kuru in New Guinea has since been completely eliminated.]

In the mid-1950s when Gajdusek first proposed his hypothesis that Kuru might be transmissible, he suggested a way to test his hypothesis. He and his colleague, C. J. Gibbs, at the National Institutes of Health in Bethesda, Maryland, would inject brain tissue from patients dying with Kuru into the brains of normal primates and other mammals and see if these animals developed signs of the disease. Meanwhile, in 1957 Dr. Igor Klatzo, a distinguished neuropathologist at NIH, had begun microscopic examinations of the brain tissue that Gajdusek had sent back from New Guinea. Klatzo discovered that the "spongiform" appearance of the diseased brains resembled that found in a rare, rapidly dementing disease in humans called Creutzfeldt-Jakob disease, so-named after its two discoverers, H. G. Creutzfeldt and A. Jakob who independently described forms of the disease in Germany in 1920 and 1921, respectively. (The incidence of Creutzfeldt-Jakob disease is less than 1 case per million people per year.)

Meanwhile, William Hadlow, an astute veterinary pathologist, who had recently left NIH's Rocky Mountain laboratory for the Agricultural Research Council Field Station at Compton in England, recognized that the recently published clinical and neuropathological findings of Kuru in humans resembled those previously described in scrapie, a rapidly fatal brain disease in sheep. Hadlow fired off a succinct one-page letter to the editor of *Lancet,* the prestigious British journal, describing the numerous and remarkable similarities between the two diseases. Hadlow also recounted the earlier successes of the veterinary community in transmitting scrapie from clinically affected sheep to previously normal sheep by inoculating small amounts of brain tissue from diseased animals into the brains of certain species of normal animals. At the time there was a printer's strike in London, so Hadlow sent off a preprint with a covering letter to Gajdusek. As it turned out, Gajdusek was away in New Guinea when his letter arrived, so it would still be several months before the NIH workers would learn of his insights. Hadlow's note to *Lancet* was finally published in the September 5, 1959 issue.

Later, when their own experiments were completed, Gajdusek and Gibbs had proof that the cause of Kuru was a novel replicating agent unrelated to any conventional virus. Moreover, the brain changes found in animals injected with the transmissible causal agent of Kuru showed pathologic changes that were indeed similar to those found in Kuru, scrapie, and Creutzfeldt-Jakob disease.

Gibbs and Gajdusek soon proved that the human Creutzfeldt-Jakob disease was also readily transmissible to higher primates just as Kuru was. In fact, pathology in the brains of the recipient animals in the cases of the two diseases was so similar that they postulated that the two diseases were caused by very similar, if not identical, infectious agents. They speculated, too, that perhaps a chance occurrence of Creutzfeldt-Jakob disease in New Guinea long ago, coupled with the tradition of ritual cannibalism which provided the opportunity for transmissibility, might explain the origin of Kuru in New Guinea.

The startling successes of Gajdusek, Gibbs, and their co-workers in proving the transmissibility of Kuru and Creutzfeldt-Jakob disease led them to wonder what other diseases, especially those of the brain, might be transmissible. Thus, the NIH group began parallel studies on multiple sclerosis, amyotrophic lateral sclerosis ("Lou Gehrig's disease"), Parkinson's disease, schizophrenia, Huntington's disease, and Alzheimer's. The NIH group had particular reasons to suspect that Alzheimer's disease might be caused by a transmissible agent. After all, plaques, which showed at least a superficial resemblance to the senile plaques of Alzheimer's disease, were often found in the brains of Kuru and Creutzfeldt-Jakob victims. Just as some cases of Alzheimer's appeared to be sporadic and others genetic, so too did some cases of Creutzfeldt-Jakob disease appear to be sporadic, whereas others, perhaps 10 to 15 percent of cases, followed a clearly genetic pattern. Perhaps, Gajdusek reasoned, the novel infectious agent could be transmitted in a hereditary manner from generation to generation. Thus, as the NIH group began these studies they were particularly interested in studying both sporadic and genetic Alzheimer's disease.

It was against this background that Ben remembers receiving a call from a Dr. P. in the early 1970s asking whether affected members of his family would be willing to join the NIH study. Dr. Terry had told Dr. P. about Ben's family. As the conversation progressed, Ben learned that Dr. P., like Charles a decade earlier, had brought his own family to Katzman and Terry's attention. Like Ben's family, those in Dr. P.'s family who would develop Alzheimer's disease experienced the onset of their symptoms between the ages of 42 and 54. As Ben and Dr. P. continued their discussions, Ben was stunned to learn that Dr. P.'s great-grandmother, like his own great-grandmother Hannah, had lived in Bobruisk, Byelorussia, in the 1840s. Unfortunately, Ben and Dr. P. could not carry their genealogies back far enough to determine whether they shared a common ancestor.

At this time, Ben's only sister was in the advanced stages of Alzheimer's disease. Ben and his sister's family had already agreed that upon her death some brain tissue should be donated for the NIH study. A small amount of tissue would then be painlessly injected into the cerebral cortex of primates and other animals. As the years progressed, Gajdusek, Joe Gibbs, Michael Alpers, David Asher, Paul Brown, and other co-workers continued to study the animals injected with variously affected brain tissues. In time, they showed that multiple sclerosis, ALS, Parkinson's disease, and schizophrenia were not transmissible. In addition, in no case was "sporadic" Alzheimer's disease found to be transmissible. However, in two cases primates apparently injected with tissue from patients dying with familial Alzheimer's disease appeared to contract a brain disease with features characteristic of the transmissible encephalopathies. (These were not brains from members of either Ben's or Dr. P.'s family.)

At the time that Gajdusek received the Nobel Prize for his work on Kuru in 1976, he and his colleagues still believed that familial Alzheimer's disease might in some cases result from transmission of a novel infectious agent. They considered such an implication so important that they were in the midst of numerous attempts to replicate these results. By 1980, in a paper by Jaap Goudsmit and colleagues, the NIH group reported that they had been unable to duplicate the transmissibility of Alzheimer's disease in any other case or even upon injection of tissue from the original two brains previously suspected of harboring an infectious agent. (The apparent transmission in the two cases has never been explained. Conceivably, there was a mixup in the animals tested, or there might have been a contaminated needle, but no one knows for sure.)

Thus, by the late 1970s Gajdusek's group had become more

and more convinced that familial Alzheimer's disease was not caused by a transmissible agent. This conclusion reinforced the conviction that familial Alzheimer's disease must truly be a genetic disorder. Consequently, Gajdusek's group would soon begin important studies on the genetics of familial Alzheimer's disease, and their efforts would focus on the families of Ben and Dr. P.

Even to the present day, the NIH group continues to follow all primates injected with tissue from brains of patients dying with familial Alzheimer's disease. If no symptoms develop, the animals are allowed to live out their natural life spans. In 1989 Dr. Paul Brown could still write that the NIH group had not yet found even a single verifiable case where inoculation of tissue from the brain of a patient with Alzheimer's had transmitted the disease to an animal.

In July 1990, when I met Dr. Gajdusek, he offered to check the NIH records to see whether the animals injected with tissue from members of Dr. P.'s and Ben's family had survived long enough to exclude a transmissible dementia. Paul Brown then searched the NIH records and told me in August 1990 that two primates injected with tissue from a member of Dr. P.'s family had lived out their life spans, dying 8 and 10 years, respectively, after injection of brain tissue, and that the animals receiving injection of brain tissue from a member of Ben's family were alive and well 14 years after the intracerebral injection. Dr. Brown assures me that these long survivals without any symptoms had never been seen in the transmissible viral dementias and provided extremely strong evidence against an infectious agent being the cause of the familial Alzheimer's disease in Ben's and Dr. P.'s family.

Following upon the work of Gibbs and Gajdusek, a major advance in the study of these novel transmissible agents occurred in the late 1970s when Pat Merz and her colleagues at the University of North Carolina identified minute fibrils in the brains of scrapie-affected mice. The fibrils, which she called scrapie associated fibrils (SAF), were not found in the brains of normal mice. Merz's group further realized that the fibrils, although amyloid-like, were not identical to the amyloid plaques found in Alzheimer's disease. Then, in 1982, Stanley Prusiner coined the word "prion" (pronounced pree-on), short for "proteinaceous infectious particles," to describe his candidate for the infectious material—a substance believed by most workers in this field to be identical to the fibrils described by Merz's group. Since then,

Prusiner's group along with numerous others have made astonishing progress toward understanding the molecular biology of "prion diseases."[3] An abnormal gene either directly "causing" the familial form of these diseases or conferring susceptibility to some as yet unidentified agent or process has been found on chromosome 20, and different mutations of this same gene have been found to produce slight disease variants in different families. Further examples of familial dementias caused by prions continue to be discovered. However, all the available evidence as reviewed by Paul Brown in 1989 suggests that neither sporadic nor familial Alzheimer's disease is caused by transmissible agents.[4]

The infectious agent or prion-causing scrapie in sheep became a matter of intense veterinary concern in the late 1980s when this agent "jumped" the "species barrier" and produced a fatal bovine spongiform encephalopathy with a pathology resembling that of scrapie and Creutzfeldt-Jakob disease. It occurred in a small fraction of cattle in England inadvertently fed food processed from scrapie-infected sheep. Most, but not all, experts in this field believe that the possibility of transmission from such infected cattle to humans is extremely remote.[5]

Base Camp, Circa 1980

Even as Gajdusek's group was becoming convinced that neither sporadic nor familial Alzheimer's disease was caused by an "unconventional" slow virus, many workers still wondered whether Alzheimer's disease could be caused by virus infections in genetically vulnerable patients. By the late 1970s several convergent lines of inquiry had made this question a readily approachable problem, and, once again, Hannah's descendants would participate in a major research effort.

Earlier, the blood group antigens had been recognized when red cells from one individual were mixed with antibodies from another. Now the surface antigens of white blood cells (lymphocytes), skin, and other body cells became identified when researchers attempted to transplant tissue from a donor to a recipient and realized that the donor's organs were rejected unless the tissues were "compatible." Collectively, these important antigens comprised in humans what was called the major histocompatibility complex, or MCH, which by the mid-1970s was known to be controlled by a cluster of genes close to each other on the short arm of chromosome 6. The human antigens that make up the MCH are also called the human lymphocyte (or leukocyte) antigens, or HLA types, and these antigens, just as the blood types, differ greatly from one individual to another.

The susceptibility of an individual to certain virus infections and autoimmune diseases, such as multiple sclerosis and rheumatoid arthritis (where antibodies initially directed against a viral invader later mistakenly attack certain normal cells of the victim), may be determined by the individual's HLA type. A given individual may be more susceptible to a virus possibly because some viral antigens "mimic" one or another of the host's HLA antigens, thus giving that virus a greater chance to penetrate the host's immune defenses. Thus, many scientists turned their attention to the problem of whether Alzheimer's disease might be associated with a given HLA subtype. Such a subtype might be associated with increased susceptibility to a viral infection affecting the brain. Alternatively, the subtype might simply serve as a marker, implying that a nearby—but functionally unrelated—gene conferred susceptibility to Alzheimer's disease.

By the late 1970s a number of research groups had attempted to determine whether Alzheimer's disease was associated with any particular HLA type, but no consistent result emerged. Gajdusek's group decided to reexamine the problem. His group would also test for linkage to other markers which were then known to be controlled by genes on chromosome 6. The researchers restricted their study to the familial form of Alzheimer's disease, where it was already clear that they would be dealing with a pure form of the disease controlled by a single gene. It was still not known whether this abnormal gene left its victim uniquely susceptible to a specific virus, which then produced Alzheimer's, or whether the basis of the genetic abnormality was unrelated to any infectious agent.

From their previous studies, in which they had tested for and failed to find transmissibility of Alzheimer's disease, Gajdusek's group had established contact with a number of families with the familial form of the disease. After screening many large families with the most classical characteristics of Alzheimer's disease, they concluded that their work would best be carried out on Ben's and Dr. P.'s families. As Gajdusek told me, "These two families had as classic a form of Alzheimer's disease as we could find. That's why we selected them."

Gajdusek chose Jaap Goudsmit, a brilliant young virologist from Holland, to lead the study and examine all affected patients and at-risk relatives to ascertain their correct neurological status.

Ben welcomed these studies and assisted the investigators in contacting members of his family. Ben's only sister had died in 1973, and two of his three brothers later fell victim to the disease.

Gajdusek's results with the families were published in 1981. Detailed descriptions of the genealogy of the two pedigrees, the clinical and pathological characteristics of the disease in Ben's and Dr. P.'s family, and some descriptions of previous work by Robert Terry and others, appeared in the paper. Dr. P.'s family became identified in the medical literature as the "M.P." family after the initials of Dr. P.'s oldest brother who had fallen victim to the disease. Hannah's descendants were cited as the "S.W." family after the initials of one of Ben's affected brothers.

Goudsmit, Gajdusek, and their colleagues failed to find a correlation between the transmission of Alzheimer's disease and the tested markers on chromosome 6. Their work, though classic in one sense, with its detailed and careful design, still pointed out the technical limitations in carrying out systematic genetic studies on diseases of unknown cause where the initial biochemical abnormality caused by the abnormal gene could not yet be identified.

In many other respects, however, the recent history of medical genetics had appeared as a succession of triumphs. Indeed, the stunning discoveries of medical genetics were now accumulating at a pace that was outstripped only by the advances in molecular genetics. The two fields—one with a clinical orientation toward helping patients and providing genetic counseling for their families and the other with an emphasis on understanding basic and molecular science of the gene and cell—were coalescing with every passing year.

A comprehensive review published in 1977 by Victor McKusick working at Johns Hopkins and Frank Ruddle at Yale titled "The Status of the Gene Map of the Human Chromosome" showed how far both fields had come in the previous 20 years. It seemed hard to believe that the correct number of human chromosomes had been determined only as recently as 1956. And, too, only in the previous decade had new methods for staining the chromosomes been developed in Copenhagen by Torbjorn Caspersson. His "banding techniques" revealed distinctive regional landmarks within each chromosome which permitted their unique identification. (See following page.)

Since Watson and Crick's stunning insights into the structure of DNA in 1953, researchers had confirmed the hypothesis that the genetic messages of DNA were conveyed by the ordered sequence of their base-pairs. Yet how could a genetic disease of unknown cause be directly linked to a defective gene hidden somewhere

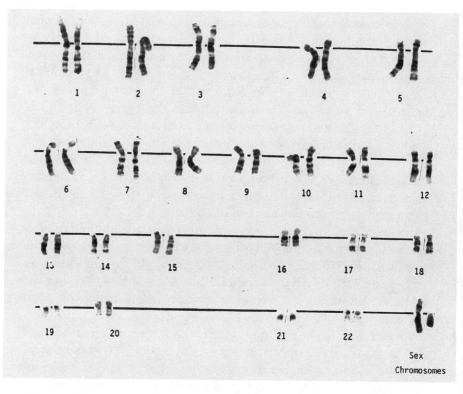

Human chromosomes. A human male "karyotype" or photomicrograph of an individual's chromosomes arranged in a standard format with Giemsa banding (G banding). The chromosome pairs are numbered 1 to 22 in standard classification, with sex chromosomes (X and Y) representing the 23rd pair. (Photomicrograph courtesy of Dr. Philip L. Townes, University of Massachusetts Medical Center.)

along DNA's 3 billion base-pairs? This remained an intractable problem. In the late 1970s when Gajdusek's group began its attempt to locate the gene for Alzheimer's disease in Hannah's descendants using the most modern techniques then available, no investigator could have foreseen the experimental and theoretical advances for linkage studies that would soon follow.

The experimental breakthroughs came first. Their essence involved the development of technologies that would enable molecular geneticists to cleave the DNA of individual chromosomes of some hundreds of millions of base-pairs into selectively cut fragments of manageable length. In December 1978 the Nobel lectures of the pioneers of these technologies—Werner Arber,[1] Hamilton Smith,[2] and Daniel Nathans[3]—in a sense brought a close to a proud age of genetics and simultaneously opened the way to a new era of vast potential.

The first glimmer in this series of monumental discoveries came in the early 1960s when Werner Arber and his colleagues at the University of Basel discovered that certain viruses called "phages" multiply within bacteria and are in fact degraded or "restricted" by certain bacteria, such as the *E. coli* bacillus, the common bacteria found within the human colon. Arber predicted the existence of bacterial "endonucleases" or "restriction enzymes" which could cleave DNA at specific sites. The restriction enzymes thus might protect a bacterium from a viral invader or on other occasions permit useful DNA from another organism to be incorporated within its own genome. (Arber also correctly predicted that bacteria would produce another type of enzyme to close the gaps in their own DNA produced by their own restriction enzymes.) The restriction enzymes that Arber and his colleagues studied actually cleaved DNA at random and were not useful enzymatic tools for DNA analysis.

Hamilton Smith and Kent Wilcox at Johns Hopkins made the chance discovery of the first site-specific restriction enzyme in 1970. Numerous such enzymes were found in various bacteria, and Smith's group proved that DNA cleavage was uniquely specific. In fact, these enzymes recognized specific sequences. For example, one enzyme from the H influenza bacteria cleaves a double strand of DNA after the first adenine (A) whenever the sequence AAGCTT is found. Another enzyme may cut between a G and a C nucleotide whenever the sequence CAGCTG is found. As Smith noted in his Nobel lecture published in 1979,

> A collection of these enzymes, each with its own particular sequence specificity, can be used to cleave DNA molecules into unique sets of fragments for DNA sequencing, chromosome analysis, gene isolation, and construction of recombinant DNA. The latter, together with the concept of molecular cloning, has given birth to the new field of genetic engineering, and from this many new and exciting medical and research applications are expected.[4,5]

One of the first to realize the significance of Smith's discovery was his Hopkins colleague, Daniel Nathans, who while on sabbatical at the Weizmann Institute in Israel received a letter from Smith describing the new discovery. Nathans realized that the genomes of DNA tumor viruses could be dissected in this way. Then after the individual fragments of cleaved viral DNA could be isolated, researchers could map out which segments of the genome were responsible for the various biological activities of the virus. The sections cut by the endonucleases were subsequently called "restriction fragments."

Nathans also recognized that if he could cleave all of the DNA into small fragments and determine the order of the fragments, then he could construct a "cleavage map" or, as it is now called, a "physical map" of the viral genome. This physical map would represent an enormous technical advance over the linkage map of the classical geneticists, for now unknown genes could be referenced to overlapping sets of cloned DNA fragments whose positions on a chromosome are known rather than to linkage markers mapped to each other only in terms of their frequencies of linkage. Nathans also knew that once physical maps of chromosomes were available, then the ultimate step in DNA mapping—the determination of the nucleotide or "sequence map"—could proceed.

In fact, soon after Nathans began to isolate the restriction fragments, the research groups headed by Sherman Weissman in New Haven and by Walter Fiers in Ghent started to analyze the nucleotide sequence—that is, the precise order of the base-pairs—within the fragments. As a result of the concurrent development of rapid DNA sequencing methods in the mid-1970s by Frederick Sanger and A. R. Coulson in Cambridge, England, and by Allan Maxam and Walter Gilbert in Cambridge, Massachusetts, the Weissman and Fiers groups were able to complete their nucleotide sequence maps of an entire monkey virus called "Simian Virus 40," or SV40 genome, by 1978. In his Nobel address of December 1978, Nathans emphasized that the same methods that had been used for decoding a viral genome and for proceeding from a genetic linkage map to a physical map, and thence to a nucleotide sequence map, could be applied in the search for the structure of mammalian genes.

Thus, if one could find a "marker" very close to an abnormal gene, then the way was open to find the location of that gene. In this way, researchers could sequence the nucleotides comprising the gene and—using additional stratagems—determine the structure of the protein expressed by a given gene, whether normal or defective, in any specific cell.

The two decades or so since the daring Charles had initiated blood group linkage studies on his own family had witnessed once unimaginable developments in medical genetics. Yet as this age closed, the techniques available for finding abnormal genes for diseases whose gene product was yet unknown had hardly changed since the turn of the century—namely, to test for linkage of the disease in a family by using the presence or absence of some kind of marker, whether a trait such as eye color or an already identified protein such as a blood group antigen.

For the molecular sciences, the twenty intervening years had been astonishing ones with momentous discoveries coming one upon another. For some of Hannah's descendants, however, the passage of the two decades must have seemed interminably slow. Still, there was much reason for hope. The contributions of medical science now stood like a vast and well-provisioned base camp from which a new generation of investigators could confidently seek higher ground. When the next steps would be taken, new objectives—like a new range of tall peaks as one ascends above the mists—would suddenly appear in clarity and in unexpected grandeur.

PART III
The Ascent

A New Basis for Linkage Maps of the Human Genome

When Robert Terry began his electron-micrographic studies of brain tissue from patients with Alzheimer's disease in the early 1960s, his initial motivation had come as a result of an option forced upon many scientists at a critical point in their career. Electron microscopy was then a powerful new technique, and Terry's research program began as a technique in search of a problem. Few neuroscientists of his period had the luxury of selecting a major problem of interest and having the appropriate methods available to solve the problem.

Over a decade and a half later, Gajdusek's group and others focused on the key problem of trying to determine if a gene causing familial Alzheimer's disease linked to a cluster of genes on chromosome 6. However, given the relatively small number of genetic markers then available, these groups would have run into a methodological stone wall if they had attempted a comprehensive linkage study testing all chromosomes. No method had yet been devised to permit a systematic search for an abnormal gene along all the vast stretches of DNA, of all the chromosomes, together comprising the 3 billion base-pairs of the human genome.

Science advances most rapidly when new technologies, crucial discoveries, and fresh ideas come together. The decisive moment for molecular biology, for example, occurred at midcentury when the development of X-ray crystallography by Sir Lawrence Bragg, the early crystallographic studies on DNA by Maurice Wilkins, and the subsequent technologically superb findings by Rosalind Franklin, coupled with the extraordinary insights of Watson and Crick, led to the discovery of the structure of DNA. Now as the late 1970s progressed, yet another revolution in molecular genetics was in the making. And when it came, the new techniques were provided not by the development of complex scientific equipment and methods like that used in electron microscopy and X-ray crystallography but rather by ingenious applications of the remarkable enzymes—the endonucleases or restriction enzymes—produced by lowly bacteria as they had evolved over hundreds of millions of years. These highly specific chemical scissors that cleave strands of DNA only at specific sequence sites had already become a cornerstone of recombinant DNA technology and a tool in determining the precise sequence of gene bases whose locations were known.

By the late 1970s normal and abnormal genes producing normal and abnormal proteins could in theory be localized and then sequenced. The central doctrine of molecular genetics held that a given sequence of DNA contained the information to specify a corresponding sequence of RNA, which in turn uniquely defines the structure of a protein. Conversely, if a scientist could determine the structure of the abnormal protein causing a given disease, he or she could deduce the sequence of the bases in the coding portion of the gene that encodes that protein. With this knowledge, the scientist could synthesize a complementary strip of DNA as a "probe" and label it with a radioactive isotope; this synthetic strip would chemically bind uniquely to the coding gene. The radioactivity emitted by the attached probe could then be traced by various methods to reveal the gene's location and identity.

Yet a vast number of human genetic diseases were caused by abnormalities either in genes where the abnormal protein was unknown or in unidentified regulatory genes that modified the expression or activity of other genes. Now the revolution begun by Arber, Smith, and Nathans, which had already led to recombinant DNA technology and to the first attempts to sequence genes of known gene product, would also open the way for the systematic search for genes of unknown function.

Since gene mappers could not study genotype (DNA sequences) directly and could only examine the co-segregation of phenotypes (observable characteristics) expressed by either normal or abnormal genes, they searched for the statistically significant co-segregation or grouping together of phenotypes within individuals in a direct line of descent through a family over several generations. Various phenotypic markers such as the red cell antigens of the ABO blood group or variably expressed serum proteins represent products of "alleles," or alternate subtypes of an expressed gene.

In the most general case, when two phenotypes are expressed by two such alleles, we can identify one phenotype or marker as "A" and the other as "B." Let us suppose that there is a genetic disease in a family and we wish to ascertain whether the gene for the disease in this family lies close to the genetic locus responsible for the A and B markers.

As we see in the diagram which follows, the disease (signified by a darkened square for males or a darkened circle for females) occurs in the grandfather who carries one allele for A and another for B. The unaffected grandmother carries two alleles for A. We find in the next generation that both a son and a daughter who inherited their father's allele for B have also inherited his disease. However, their sister who inherited her father's allele for A is unaffected. When we review the next generation, we find that the son who received the allele for B from his affected mother has also inherited the disease, whereas his sister who inherited her mother's allele for A has escaped the disease.

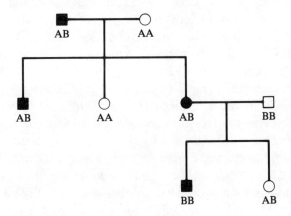

Thus, in this family the genetic defect may be carried at or close to the genetic locus for the B allele. However, since such a small number of individuals are involved, chance may explain the coincident grouping of the disease and the allele for B. Before there can be statistically robust evidence for or against linkage, consequences of many more matings in this family over more generations must be examined.

In this family the markers were "informative" in the sense that the affected grandfather carried two different alleles, A and B, the latter quite possibly linked to the disease. The unaffected grandmother expressed an AA phenotype, making it easy for the gene mapper to discern a pattern that the genes for the disease and for the B marker co-segregated. However, if the grandfather had carried two alleles for A, his mating with a spouse also carrying AA would have been "noninformative," for all the offspring, affected and unaffected alike, would have expressed an AA phenotype.

Gene mappers, at the mercy of the relatively small number of phenotypic markers available for linkage studies, had to work with whatever allelic variations and marker systems had evolved over the ages. They could not create new phenotypic markers at will. Moreover, even the successful linkage of a genetic disease to a marker left geneticists far from their ultimate goal of finding the defective gene and sequencing its DNA.

The big breakthrough that took us from a small number of phenotypic markers to an abundant number of genotypic markers—and direct access to the genes—came in 1974. This was the realization that variations, or "polymorphisms," in the *length* of the restriction fragments of the DNA cleaved by the restriction enzymes of a bacteria could be used as markers in genetic linkage studies. In 1974 T. Grodzicker and his colleagues used differences in the length of restriction fragments to localize the site of temperature-sensitive mutations in a virus. That same year, Clyde Hutchison and colleagues found length differences in restriction fragments of DNA in the mitochondria of cells from two different humans. (Mitochondria are organelles in the cytoplasm of cells that function in energy production.) Within the next several years three groups working on yeast had discovered length variations for the cleaved fragments of specific genes and used them as markers in their analysis of DNA in yeast. At this point, these workers and yeast geneticists David Botstein at MIT and Ronald Davis at Stanford could not have guessed the significance their own work and that of their colleagues Howard Goodman, Maynard Olson, and Benjamin Hall at the University of

Washington would hold for human genetics within a few short years.

Meanwhile, a major technical advance occurred in 1975. The Scottish scientist Edwin M. Southern, while on sabbatical leave in Zurich, developed a sensitive new method for cleanly separating restriction fragments according to length. In this method known as "gel electrophoresis," the fragments are set out on a gel atop a plexiglass plate, and an electric current is passed between two electrodes—one above, and the other well below the region where the still unseparated fragments are placed. The rate of migration of the electrically charged fragments in response to the electric current is determined by their size and resistance to movement through the gel. The shorter fragments, for example, will move faster through the gel than the longer fragments. After the DNA fragments have been separated, they are lifted off the gel with a piece of blotting paper in order to preserve their relative positions. When radioactively labeled "complementary" DNA probes are introduced, the probes bind only to those restriction fragments that contain the complementary DNA sequence. The position of the sought-after restriction fragment, which is also a measure of its length, is determined when the localized radioactivity exposes an overlying X-ray film. The sequence of bases in an isolated fragment can then be determined using other methods. This procedure has been aptly named a "Southern blot" after its developer.

By 1978, Southern's technique facilitated the first clinical application of restriction fragments. Yuet Wai Kan and Andree Dozy at San Francisco General Hospital carried out an analysis of restriction fragments in a human globin gene and discovered a fragment of one length in people of either Caucasian or Asian ancestry, and two variants of different lengths in some people of African origin.[1] When a person of African ancestry was found to have received two copies of the longer fragment, a 13,000 base variant (abbreviated 13.0 kilobases or kb), having received one copy from each parent, that individual was frequently discovered to be afflicted with sickle cell anemia.

At some point in the history of Africa, perhaps when the malaria parasite reached native populations, people receiving a single copy of the gene carrying the sickle cell mutation were protected against malaria. The mutation which expressed itself in one-half of the hemoglobin molecules protected the red blood cell against invasion by the malarial parasite. Thus, the chance that the descendants of these carriers of one sickle cell gene would survive increased, which led to a large increase in the frequency of the gene in that population. Unfortunately, a person receiving two copies of the gene does develop sickle cell anemia, a frequently life-threatening illness. (Sickle cell anemia affects about 1 in 400 Americans of African ancestry.)

Kan and Dozy were particularly interested in the prenatal diagnosis of sickle cell anemia, but at that time the diagnostic method used to obtain a sample of fetal blood posed high risk to the patient. Kan and Dozy reasoned that a prenatal diagnosis of the disease might be obtained more readily and more safely if fetal cells were taken for analysis of restriction fragments of DNA from the mother's amniotic fluid—the fluid within the intra-uterine sac confining the embryo.

In papers published in October and November 1978, Kan and Dozy realized that they had discovered much more than just another prenatal diagnostic test: "Restriction endonuclease mapping may provide a new class of genetic markers for linkage studies."[2] They foresaw, too, the generality of their method applied to the study of other hereditary diseases in which a structural gene (i.e., a gene encoding a protein) had already been identified.

The importance of Kan and Dozy's papers was immediately recognized. Ellen Solomon and Walter Bodmer at the University of Oxford in England wrote a single-page letter to the editor of *Lancet*, published in the April 29, 1979 issue, bearing the unassuming title "Evolution of Sickle Cell Variant." The original purpose of their note had been to offer an alternative date for the origin of the sickle cell gene in the human population. However, in the course of their analysis of the evolution of the sickle cell mutation and the restriction fragments, Solomon and Bodmer suddenly conceived a brilliant idea of great consequence. Their letter closed with the following prophetic words:

> As mentioned by Kurnit, polymorphisms for restriction enzyme sites are likely to provide powerful new tools for evolutionary and genetic studies. These polymorphisms, which identify variation at the level of the gene itself, may provide a quantum jump in the range of available genetic markers for study. Kan and Dozy have emphasized their application to the general problem of prenatal prognosis of hereditary diseases, even where the specific biochemical defect is not known. If the level of polymorphism for these sites is as high as the initial studies suggest, and given the range of available restriction enzymes, one can envisage finding enough markers to cover systematically the whole human genome. Thus, only 200–300 suitably selected probes might be needed to provide a genetic marker for, say, every 10%

recombination. Such a set of genetic markers could revolutionize our ability to study the genetic determination of complex attributes and to follow the inheritance of traits that are so far difficult or impossible to study at the cellular level.[3]

In just half a dozen sentences, Solomon and Bodmer had appreciated that the new markers could be used to map unknown genes. They also estimated that as few as several hundred suitably placed markers would be enough to localize a genetic defect onto a given chromosome within about 10 million base-pairs. During meiosis, the process of generation of the sperm and ovum, there is on average about a 10 percent chance of a chromosomal recombinant event—that is, "crossover" between genes roughly 10 million base-pairs apart. Conversely, there is a 90 percent chance that two genes spaced within 10 million base-pairs of each other will be inherited together by members of the next generation.[4] Thus, as a first step toward linking a disease to a chromosome, gene mappers would need only sample every 20 million base-pairs over the genome, and then no abnormal gene would be more than 10 million base-pairs away from at least one marker. Essentially, Solomon and Bodmer reduced the problem of locating a genetic variant among some 3 billion base-pairs to a feasible first step of linking a trait or disease to one or only several adjacent DNA markers from a pool of only several hundred.

Concurrently, in England at the University of Leicester Alec Jeffreys's primary motivation was "to estimate the overall degree of genetic diversity in the human genome" by determining the frequency of normal variations in human DNA sequences. By now Mary-Claire King and Allan Wilson[5] at Berkeley had discovered that amino acid sequences in proteins in humans and their nearest ancestors among the great apes differed by only about 1 percent. Assuredly, differences in normal protein structure between members of the same species must be much less. Jeffreys set out to estimate variance between humans, not at the protein level, but at the level of their DNA. Choosing to study sequence variations in several distinct globin genes, he analyzed relevant DNA sequences from 60 unrelated individuals. He performed the analysis by separately digesting or cutting the DNA using one or another of the restriction enzymes. After many months of effort, and with no prior guarantee of success, he discovered three types of restriction fragments of varying length in normal individuals.

For several years, scientists had known that most genes consist of coding regions of DNA (now called exons) which determine the precise sequence of amino acids expressed in the protein specified by a given gene. Within a gene the exons are separated from each other by noncoding intervening sequences (now called introns). A given gene can in fact direct the synthesis of proteins differing in composition, depending on how strips of messenger RNA, specified by their respective exons, are "spliced" as a prelude to protein synthesis. Successive genes on a chromosome are also often separated from each other by long tracts of noncoding DNA for which the function—if any—is presently unknown. Thus, Jeffreys tried to discern the location of his sequence variants. In time, he found that all three of his variants lay within the noncoding intervening sequences or introns within a structural gene. Within his group of subjects, the DNA sequences encoding the normal proteins were identical.

From his data and various assumptions, Jeffreys estimated that, on average, about 1 in 100 base-pairs along the genome will vary in humans. Indeed, diversity is the very essence of genetics. It is this diversity that accounts for both the rich differences between normal individuals and the tragic consequences of a genetic disease when the mutation produces a defective gene. Jeffreys realized that any inheritable variant—whether blood type or the length of a restriction fragment of DNA—could be used as a marker to test for linkage to any other DNA fragment trait or genetic disease. In his words, "DNA sequence variants would serve as useful new markers in examining the population structure and origin of human races, and already have important implications in the detection and analysis of genetic disease."[6]

Jeffreys concluded his paper on DNA sequence variants, published in 1979, with a prediction: "With banks of human DNA clones now available . . . to provide probes into all parts of the human genome, new polymorphic variants should be detectable on all human chromosomes and therefore ultimately be used for the indirect determination by linkage of any genetic disease suitably linked to a restriction enzyme cleavage site polymorphism."[7] Thus, starting with a scientific objective that was seemingly remote from that of mapping genetic diseases, Jeffreys had struck upon the idea of using variations in the length of the DNA fragments cut by bacterial enzymes as the basis for a new class of probes that could be used to map the human genome.

Solomon and Bodmer's germinal ideas were published in April 1979; Jeffreys's findings came out in September of that year; and Bodmer's own detailed elucidation of concepts and methods were delivered in his Allen Memorial Address in September 1980. Any of these alone could have triggered a new revolution in molecular genetics. However, at the time they were preparing their own papers, none of them knew that a remarkable event had already occurred a full year previously in April 1978, even several months before Kan and Dozy's two papers had been published. This event would even more swiftly change the course of human molecular genetics.

In April 1978, David Botstein and Ronald Davis were invited as distinguished experts and commentators to a retreat for graduate students in biology and genetics enrolled at the University of Utah; the retreat was held at a lodge in Alta, Utah, high in the Wasatch Mountains overlooking Salt Lake City.[8] On the second day of the conference, Kerry Kravitz, a graduate student of Mark Skolnick, presented his work on how the disease known as hemochromatosis was inherited (i.e., whether the inheritance was dominant or recessive). Workers in France had recently linked this disease to the HLA complex now known to reside on chromosome 6. Neither Botstein nor Davis had any previous major interest in human genetics or in linking specific diseases to genetic markers. Skolnick, a population geneticist by background, then discussed Kravitz's presentation and made clear to Botstein and Davis the importance of discovering previously hidden genes by their linkage to known markers. At this point, another of Skolnick's graduate students, Jon Hill, commented on his group's difficult search for a gene-conferring susceptibility to breast cancer: his sought-after gene was not linked to one or another of the then known polymorphic set of markers like the ABO blood group markers or those of the HLA complex. His remark energized Botstein and Davis to link their own experience using bacterial enzymes to cleave DNA in yeast with the problem at hand.

Jerry Bishop and Michael Waldholz in their book *Genome* interviewed the participants at the Alta meeting and tried to reconstruct the sequence of events.

> I remember there was this long silence in which Davis and Botstein just looked at one another," Skolnick says. "Then they suddenly started talking about transposons, Southern blots, deletions, restriction enzymes. Finally, Botstein looked at me and said something like, 'Theoretically, we can solve your problem, we can give you markers, probably markers spread all over the genome. Nobody has specifically looked for them or [when they saw them] thought about them as anything but junk. But they exist. It's not proven, but I think they exist over all the chromosomes. Moreover, they can be identified and their locations on the chromosomes pinpointed.'"[9]

Up until this time the restriction fragments of variable length, which Botstein and his colleagues would henceforth call restriction fragment length polymorphisms, or RFLPs, had been used in genetic linkage analysis only *after* prior isolation of the DNA of a *known* gene of interest. Such had been the case in Botstein and Davis's independent studies in yeast, and in the work on the human β-globin gene by Kan and Dozy, which would soon be published.

At the Alta meeting, Botstein and Davis had suddenly grasped, as they later wrote, that "for the more general purpose of mapping genetic loci, RFLPs need not encode the gene of interest, but only be sufficiently nearby to display genetic linkage. This is extremely important, because a recombinant DNA probe need not reveal a restriction fragment containing part of the gene of interest to be useful; that is, one does not have to 'isolate the gene' to map it."[10]

Moreover, they realized that the occasional variabilities in DNA sequence found in and near the region of known genes might in fact be scattered more or less randomly all over the entire genome. Thus, the short strands of DNA cut by the bacterial enzymes would themselves show variations or polymorphisms and thus be useful as markers.

By the time they submitted their paper, Kan and Dozy's work on the sickle cell variant, Jeffreys's demonstration of hereditable and detectable RFLPs at the normal β-globin locus, and Tom Maniatis and colleagues' finding of a polymorphism in an intervening sequence of the β-globin gene had all appeared. The published work strengthened Botstein and his colleagues' assumption that such hereditable RFLPs would exist scattered throughout the genome. They believed that the previous work "implies that the genetic mapping scheme based on linkage in family studies of randomly derived RFLPs is possible in principle." But they cautiously noted: "It is less clear, however, how difficult it will be to find useful polymorphic loci and to establish linkage relationships in practice."[11] In other words, they still did not know whether they could find markers that would identify unique sequences of DNA at a single location; a marker that bound indiscriminately to numerous strips of DNA on multiple chromosomes would be useless to pin down the location of a single gene. Nor did they know whether enough useful markers could be found to scan the entire genome at intervals that were reasonably spaced.

Botstein and his group quickly isolated the hard issues that would have to be resolved if their scheme was to work: "We must answer a series of questions many of which can have independent and complex answers. Four interrelated questions must be answered: (1) How many markers are needed? (2) How polymorphic must each marker be? (3) How many families are needed to establish linkage? (4) How much polymorphism can we expect in the human genome?"[12]

As chance would have it, Botstein and Davis had found in Skolnick a highly skilled population geneticist who already had the background in human genetics and in statistics to address these questions with them. Based on work by others, they estimated that variations in DNA sequence would occur at least once every 1,000 base-pairs. This likely spacing interval suggested that enough variability would exist to enable the entire human genome to be mapped with only several hundred polymorphic markers. However, there was no certainty that Botstein and Davis were correct. Some critics would later argue that too few variations would be found for RFLPs to be generally useful as genetic markers.

If Botstein and Davis were correct, however, then any genetic disease could henceforth be mapped to one or another of the RFLPs. They estimated, just as Solomon and Bodmer had, that useful RFLPs and the sought-after genes would have to be within 10 million base-pairs of each other so that the RFLP and the target gene had a 90 percent chance of remaining together on the same chromosome during the gene shuffling that occurs at each meiosis when the germ cells are formed. Based on this calculation, they estimated, as had Solomon and Bodmer, that they would need about 150 RFLPs scattered roughly every 20 million bases along the entire human genome.

Both Botstein's and Solomon and Bodmer's groups were overly optimistic on the low number of only 150 RFLPs that would be needed. Their estimate assumed that the matings within a family would be highly informative from a genetic point of view (as in the example given for the small family tree earlier in this chapter), but such informativeness for a given set of markers seldom holds throughout an extended pedigree. Moreover, both groups somewhat underestimated the size and number of families that would be needed to detect linkage. Even so, the general principles for using RFLPs for markers were so sound and the opportunities for advance so great that a revolution would soon be in the making.

Before gene mappers could systematically test for linkage of a genetic disease to a set of markers, they had to construct linkage maps of the novel RFLP probes that could relate the proximity of two or more of these markers and give their order along a chromosome based on estimates of the probability that the members of a given pair of markers are inherited together from a parent. To create such maps, geneticists would need blood samples from members of large multigenerational families whose ancestry was well documented. In the early 1970s the Mormon church had given geneticists at the University of Utah access to their extraordinarily accurate genealogical records. These records, coupled with the Mormons' traditionally large families, provided geneticists with an invaluable resource. But it was not mere chance that the Mormons had established such meticulously accurate records. The Mormon elders had established such detailed geneologies so that they could confer vicarious baptisms for the salvation of the dead on the deceased ancestors of living church members and upon others who they believe would have accepted their church's teachings had they been able to have learned of these doctrines during their lifetimes.

Skolnick was in a superb position to assist in the mapping and statistical work since he had initially come to Utah to organize the Mormon genealogical records and apply computer programs that he had developed for his own research projects. One of Skolnick's original interests was to discover genetic factors that confer susceptibility to the development of various cancers. It would not be long before Skolnick became a leader in this field.

Meanwhile, someone would have to confirm whether Botstein and Davis were correct when they claimed that polymorphic strips of DNA (the RFLPs) were scattered throughout the human genome. Both workers had full research programs under way in their respective laboratories, and neither was about to launch this search. In the exciting days that followed the Alta meeting, Botstein, Davis, and Skolnick discussed the idea openly with many colleagues. Skolnick met Maurice Fox (Botstein's one-time research advisor at MIT) at a meeting at the National Institutes of Health and discussed the idea with him. Upon his return to Cambridge, Massachusetts, Fox discussed the theory with Botstein in more detail. According to Bishop and Waldholz, Fox then contacted another of his former students, an exceedingly able recent graduate student named Raymond White who was working at the nearby University of Massachusetts Medical Center. Fox suggested that White contact Botstein immediately and commit himself to searching for the postulated RFLPs. White decided to take up the challenge.

Over the next year and a half Botstein, White, Skolnick, and Davis fully documented their ideas and prepared a detailed manuscript. Their now famous paper, "Construction of a Genetic Linkage Map in Man Using Restriction Fragment Length Polymorphisms," was published in the *American Journal of Human Genetics* in October 1980.[13]

Meanwhile, Raymond White and Arlene Wyman, a postdoctoral student in White's laboratory, began the search for the RFLPs. So that their probes would have specificity for only a single segment of DNA, White and Wyman had to devise a method to eliminate all stretches of DNA containing "repetitious segments"—the short similar sequences about 250 bases long scattered throughout the genome. A year and a half after the Alta meeting and nine months before Botstein, White, Skolnick, and Davis's paper was published, Wyman and White had found their first anonymous RFLP and had developed methods that would in time yield many others. Wyman and White's paper, bearing the modest title "A Highly Polymorphic Locus in Human DNA," appeared in November 1980.[14] It began: "A locus in the human genome, not associated with any specific gene, has been found to be a site of a restriction fragment length polymorphism." Thus, the work documenting the discovery of the first of the many hundred RFLPs that would be needed to scan the human genome was now published.

But in the years before their work was published, Botstein and his colleagues had been unusually open in discussing their ideas. As a result, other geneticists did not have to wait for their publications to begin their own efforts based on the new notions. Even so, no one yet knew how well the newly conceived methods would work in practice, nor did anyone know whether the time course for discovery of an unknown human gene would be reckoned within a decade or within a half-century. Even as the galley proofs of the papers by Botstein's group and by Wyman and White slowly made their way through the lengthy process of publication, an effort was actually being planned to find an unknown gene—the Huntington's disease gene—based on these methods. The results would soon startle the world of science and medicine.

The Quest for the Huntington's Disease Gene

The decisive event that would in time trigger the systematic search for the genes causing familial Alzheimer's disease was the momentous discovery in the early 1980s that linked Huntington's disease to chromosome 4. This discovery at once proved the general applicability of the ideas of Botstein's, Jeffreys's, Solomon and Bodmer's, and of Kan and Dozy's groups. The search for the Huntington's disease gene was to have both immediate and long-term effects on the search for the genes that cause familial Alzheimer's disease.

Huntington's disease, or Huntington's chorea, is characterized by the progressive loss of voluntary motor control, with correspondingly uncontrollable choreiform or jerky movements, personality changes, depression often leading to attempted suicide, dementia, and eventually complete incapacitation leading to death, usually within 15 years after onset. The disease, affecting about 1 in 25,000 people, was first formally described by George Huntington, a Long Island physician, in 1872 in a classic paper growing out of observations that he and his physician-father had made on affected individuals from a number of families in the Long Island area.

P. R. Vessie of Greenwich, Connecticut, has traced the history of the disorder in North America as far back as 1630 when a group of settlers from England, including John Winthrop who later became governor of the Massachusetts Bay Colony, landed at Salem, Massachusetts. Two affected brothers were passengers, and their progeny spread throughout New England; some migrated to Long Island and probably gave rise to the families later studied by George Huntington. Other descendants remained in the Salem area. Because of their marked abnormalities of movement, many of these unfortunate individuals were accused and hanged as witches in the early colonial period.

In Huntington's disease, just as in familial Alzheimer's disease, many of the primary steps in seeking a cause and then a cure came from the efforts of the families actually involved. The tragedy of Huntington's disease probably first became known to the American public when it struck the great American folk singer Woody Guthrie, who came to prominence during the depression days of the 1930s. Guthrie's widow, Marjorie, founded a commission to combat Huntington's chorea during the mid-1960s. Her intention was to organize chapters around the country with the objective of raising enough money to convince Congress to allocate funds for research on Huntington's disease.

In 1968 a tragedy that befell another American family served as the springboard for an expanded attack on Huntington's disease. At 60, Milton Wexler, a distinguished psychoanalyst in Hollywood, California, could not have imaged what lay ahead for him and his family. He was already an accomplished lawyer on Wall Street in 1939 when he decided to obtain a doctorate degree and become a psychoanalyst. Later he carried out research on schizophrenia at the Menninger Foundation in Topeka, Kansas. Milton's wife, Leonore, had known that her father, a Russian-Jewish immigrant, had died of Huntington's disease at age 55 when she was only 13.[1] Later she looked the disease up in a text which incorrectly stated that Huntington's only affected males.

According to accounts by the Wexler family,[2] Leonore learned in 1950 that her 48-year-old brother, Jesse, had developed a neurological disorder. The neurologist who examined Jesse asked to see his two brothers and then concluded that all three were suffering from a degenerative brain disease that was eventually diagnosed as

Huntington's disease. With a new understanding of the genetics of Huntington's disease, Leonore realized that she and possibly her two very young daughters were also at risk.

That same year, Wexler decided to leave behind his research at the Menninger Foundation and move to Los Angeles to enter private practice in psychoanalysis so that he could assume the financial burden of supporting his brothers-in-law and plan for the contingency of illness in his own family. For a time the Wexlers thrived in Los Angeles, but then in 1964 the couple separated and subsequently divorced. At this time, Milton believed that Leonore had escaped the Huntington's disease gene because she was still asymptomatic and some years older than her brothers were when they were diagnosed with Huntington's.

According to an account by Waldholz and Bishop in their book *Genome*,[2] Leonore, while on her way to jury duty one day in 1968, was stopped by a policeman who admonished her, "Aren't you ashamed of drinking so early in the morning." Leonore, alarmed that she had been walking erratically, phoned Milton who arranged for a neurologist to see her. All too soon, the neurologist confirmed the diagnosis of Huntington's disease. " 'It shocked me,' Milton Wexler says. 'All I felt was horror. All I could think about was my two beautiful daughters, both working on their doctorates, and I got sick.' "[3]

Five days later the younger daughter, Nancy, then 24, was vacationing in France when she received a call from her father asking her to return home on the pretext of his sixtieth birthday. Nancy's sister, Alice, also in her twenties, was called home too.

Nancy Wexler has recalled the moment she learned of her mother's diagnosis:

> As gently as he could my father finally told us the truth about the secret so long hidden in my mother's family. Mother was dying of Huntington's disease. It was genetic; Alice and I each had a 50-50 risk of having inherited it. There was no test to determine if we had the gene—only time would tell. If we had it, we could pass it on to our children.
>
> I remember very little of the conversation. Just that my mother was dying and that I decided I should not have children. I know now that my father was terrified that Alice and I would give up our studies in the face of this monumental uncertainty. He says we told him that a 50 percent chance to be healthy wasn't so bad. We don't remember.[4]

According to Nancy, her father's response to calamity was to over-come it no matter what it took. At age 60, Milton Wexler took up the fight against Huntington's.

At first, Milton Wexler decided to set up a California chapter of Marjorie Guthrie's organization and to enlist the help of his friends in the motion picture industry to raise money to tackle the disease.[5] As he looked around and realized how little was known about the cause of Huntington's, Wexler realized that getting Congress to appropriate research funds would not be enough. He would have to cultivate the interest of both established scientists and new researchers just getting started if he were to make a dent in Huntington's devastating scourge.

For a time, Wexler and Guthrie worked together on a nation-wide campaign. But after Wexler helped organize a fundraising concert that included folk singers Pete Seeger, Joan Baez, and Woody Guthrie's son, Arlo, Wexler and Guthrie disagreed over the allocation of the very limited resources. Marjorie Guthrie, as always, wanted to use the limited funds as the fulcrum to lobby Congress, whereas Wexler wanted to create workshops on Huntington's disease to entice young scientists to work on the disease. As a result of their disagreement, Wexler decided to set up his own foundation, the Hereditary Disease Foundation; however, the two continued to work together to lobby Congress for research funds.

Wexler's board of scientific advisors included prominent geneticist Seymour Benzer and the biochemist William Dreyer, both at the California Institute of Technology, as well as other prominent scientists. Wexler hired a young Cal-Tech graduate student, Ronald Konopka, to visit laboratories around the country and identify young emerging scientists who might be convinced to work on Huntington's disease. Concurrently, Wexler organized his earliest workshops in Los Angeles. In parties that followed, he brought together the research scientists and his prominent Holly-wood friends, who helped raise money for research and, by their presence, encouraged scientists about the importance of their goal. Such entertainment stars as Carol Burnett, Candice Bergen, Gregory Peck, Jennifer Jones, Julie Andrews, and Cary Grant, all joined in as did many of their spouses, some equally prominent.

An even greater motivation for the research soon became evident. As Bishop and Waldholz have written,

> The Wexlers also brought something else to the foundation that sci-entists soon found captivating and irresistible—Nancy. From the

beginning, the Wexlers would introduce the scientists, most of whom confined their studies to microscopes, slides and tissue specimens, to people affected by the disease. The result was often riveting but when the scientists learned that Nancy, who was about their age and also pursuing a career in science, was at risk, the impact was often even greater.[6]

By now the young Nancy had already graduated from Radcliffe College, spent a year under a Fulbright scholarship studying the mental health problems of the poor at the University of the West Indies in Jamaica, and studied at a psychoanalytic training institute in London, under Anna Freud, the daughter of Sigmund Freud.[7] Back in the United States, Nancy began doctoral work in clinical psychology at the University of Michigan and, for a time, she was not directly involved in her father's work. Then, in 1970, after a failed suicide attempt by her mother, Nancy realized not only how horrible her mother's life had become but also how much worse her mother's future would be. Milton had not yet organized his own foundation, and Nancy now organized a Michigan chapter of Marjorie Guthrie's organization. As she began to counsel and help families affected by Huntington's disease and become totally immersed in Huntington's disease research, she decided to write her doctoral thesis on the impact of the disease on the affected families. And as her commitment to this effort grew, so did her own conviction and personal strength.

In 1972, her father's foundation organized a workshop in Ohio as part of a meeting sponsored by the World Federation in the Neurology of Huntington's Chorea. During the workshop, an event occurred which would change the course of Nancy's life and the future of research on Huntington's disease. A Venezuelan physician, Ramon Avila Giron,[8] recounted the results of his years of doctoral work documenting an unusually large number of victims of Huntington's disease, which he claimed all belonged to the same family living in several remote villages on the shores of Lake Maracaibo on the northern coast of South America.[9] Avila Giron then showed a black-and-white movie of the villagers, and the audience was astonished to see numerous members of a single community afflicted by the disease.

Avila Giron had been a student of Americo Negrette, who had served a rural internship in San Louis, Venezuela, in the years following World War II. Even though Negrette had only a rudimentary knowledge of genetics, he mapped out the pedigree and realized that the disease, which he recognized as Huntington's disease, was inherited as an autosomal dominant gene. Negrette published his work in 1955. His findings subsequently drew attention to a region where Venezuelan medical students could conduct research in genetics.

Following Avila Giron's presentation in 1972, the Hereditary Disease Foundation decided that a team ought to go to Venezuela to verify Giron's claim of the enormous incidence of Huntington's disease in a dozen or so of the small villages along Lake Maracaibo. At the time, the idea was not pursued further, but Nancy never forgot the potential significance of the Venezuelan work.

In 1976 Congress recognized Huntington's disease as worthy of federal funding and created the Huntington's Disease Commission. Marjorie Guthrie was appointed chairman, Milton Wexler the vice-chairman for research, and Nancy Wexler, the deputy director. Within several months, the director quit and Nancy, on her thirtieth birthday, was named executive director.

As the new commission searched for promising research ideas, they learned of the discoveries of Michael Brown and Joseph Goldstein in Texas who were studying the genetic defect that caused familial hypercholesterolemia. This disease is characterized by very high levels of serum cholesterol which, if untreated, can lead to severe vascular disease and heart attacks by the middle forties or fifties, or even earlier. Children who received two copies of the defective gene (homozygotes) had symptoms twice as severe as those of their parents with one copy (heterozygotes). Learning this, members of the Huntington's Disease Commission jumped at the possibility that they might identify the primary biochemical abnormality in Huntington's disease by discovering individuals who had two copies of the defective gene. Perhaps these affected individuals would show such abnormally high levels of some substance in the brain that the basic abnormality in Huntington's could be recognized.

Nancy and some of her colleagues recalled Avila Giron's presentation and realized that, if there was a homozygote anywhere in the world, that person would most likely be living along Lake Maracaibo. Nancy Wexler made her first visit to Lake Maracaibo in July 1979. By this time, there were almost 100 living people affected by Huntington's disease in the area.

After much effort and physical hardship imposed by the oppressive heat, Nancy and her colleagues discovered a family where both parents had been struck with Huntington's disease.

The couple had 14 children. On a statistical basis, one-quarter of these children would escape, receiving neither parent's copy of the abnormal gene, another half would receive the defective gene from one parent or the other, and the remaining three or four children would be expected to receive two copies of the defective gene. Yet the children were still young, and it might be many years before the presumed homozygotes developed the symptoms. Moreover, as Nancy's team realized, there was always the possibility that embryos receiving two copies of the defected gene might never reach a viable state and die *in utero* before being born. In fact, the Venezuelan mother had already had several miscarriages. Thus, after all their efforts, Nancy's team could not be sure that a homozygote for Huntington's disease even existed. Nor could they know, even if one did exist, when and how they would be able to identify such a homozygous individual.

Before Nancy left for Venezuela, another sequence of events initiated by her father's foundation was to prove of even more fundamental importance. Allen Tobin, a molecular geneticist, who by training had worked on globin genes, became the research director of the Hereditary Disease Foundation in 1978. Tobin told me that Milton Wexler had promised him that Tobin could plan anything he wanted for his first workshop, which would be held in October 1979. Tobin knew about Kan and Dozy's work on the variants in the sickle cell gene and had up-to-date knowledge of the efforts being made to sequence genes of known location.

Tobin had a general idea that the time was right for an approach to Huntington's disease that would involve molecular genetics, but he had no idea of the forces he would soon unleash. In February 1979 Tobin phoned his old friend David Housman at MIT, who also had a keen interest in globin genes; the two had been friends since the days when Tobin had worked at Harvard. Tobin invited Housman to lecture at UCLA. Housman, whose laboratory was down the hall from David Botstein's laboratory, filled Tobin in on the ideas Botstein and Ronald Davis had spawned at the meeting in Alta the previous April. Tobin realized that the search for the homozygote was now of secondary importance; the key objective was to explore Botstein and Davis's ideas further.

Thus, Tobin invited Botstein, White, and Housman to the workshop held in October 1979 at the National Institutes of Health. As Botstein and White explained their ideas and emphasized the importance of studying a gene in a large family, Nancy Wexler immediately recalled the Venezuelan kindred living on the

Dr. Nancy Wexler. (Courtesy of Dr. Wexler.)

Dr. David Housman. (Photograph by Donna Coveney, Massachusetts Institute of Technology; courtesy of Dr. Housman.)

Dr. James Gusella. (Courtesy of Dr. Gusella.)

shores of Lake Maracaibo and brought the information to the attention of the workshop participants. But at the time, Botstein and White felt that it was premature to begin a search for the genetic defect in Huntington's disease. There was only one known anonymous RFLP, the one that White and Wyman had recently discovered. It was still not clear how many RFLPs might exist, and it was still not proved that geneticists could link the RFLP probes with familial diseases.

At that time Botstein and White believed that it was far more important to isolate RFLPs on every chromosome so that geneticists would have a rough map of markers spanning the entire human genome. Botstein and White believed that a search for any specific disease gene might be futile until a full set of markers were available. The Botstein-White objective, first proposed at a conference held in the summer of 1979 at the Banbury Center associated with the Cold Springs Harbor laboratory now run by James Watson, was in essence the germ of what would evolve into the human genome project.

David Housman disagreed that the search for the Huntington's disease gene should await the availability of a full set of RFLPs spanning all the chromosomes of the human genome. He had gone to the October workshop with a plan to begin an assault on the Huntington's gene. Housman had not yet met Nancy Wexler and so did not yet know about the Venezuelan family, but he had been developing ideas on other methods to generate RFLPs as test probes.

As Bishop and Waldholz recall, Housman's reaction to Nancy Wexler's presentation on the Venezuelan family was clear: "As far as I was concerned, everything had fallen into place. We had a strategy, we were going to get funding, and now we had a family."[10]

In one sense, Housman had more faith in the immediate application of Botstein's strategy than had Botstein himself. Housman assured Nancy Wexler that Botstein's ideas would work and that she should begin collecting blood samples from the Venezuelan pedigree. Housman could not predict whether it would take 5, 10, or even 50 years, but for Nancy the idea that something would eventually work was exhilarating. Nancy also encouraged Housman to apply to her father's foundation for seed money to get the initial research started, even before his intended grant application to NIH could be submitted and hopefully approved and funded.

About the same time, another astonishing series of unanticipated events occurred that would lead within weeks to the completion of arrangements for the collaborative effort between Wexler and Housman. Housman told the story to me. During the summer of 1978, Jody Rich, a young student and son of MIT's distinguished molecular biologist Alexander Rich, had worked in Housman's laboratory. Jody spent the next summer with his family at Woods Hole, Massachusetts, the summer research home for many biologists, especially geneticists. In late August 1979 Jody sustained a head injury during a soccer game and for a time experienced diplopia or double vision. Alexander Rich drove Jody back to Boston to be examined by Joseph Martin, who had recently arrived from Montreal to become chief of neurology at Massachusetts General Hospital.

After assuring father and son that Jody's injury was of no concern, Martin mentioned his own interest in Huntington's disease. Martin had been reading Judson's then recently published masterpiece on the birth of molecular biology, *The Eighth Day of Creation*, and instinctively sensed that the time might be right for some kind of molecular approach to Huntington's disease. Martin asked the senior Rich if he had any suggestions. Rich told Martin that he'd keep his query in mind. Within days Rich ran into Housman in the corridors of MIT where Housman excitedly told Rich of his ideas to study Huntington's. As a result of this fortuitous and timely encounter, Housman learned of Joe Martin's own interest in Huntington's disease.

For Housman almost everything had now fallen into place. A collaboration was forged between Housman's MIT group and the clinicians and research workers at Massachusetts General Hospital (MGH). Martin's group had already been preparing a grant application on Huntington's disease that had to be submitted by November 1, 1979. With only weeks left before the deadline, Housman readied his contribution to the grant application.

Housman had neither the time nor lab space at MIT to carry out the work himself. Hence, the molecular neurogenetics lab would be set up at MGH, and Housman would find a young molecular geneticist who could commit his full time to the project. Housman knew just the person to carry out the hands-on genetic work. James Gusella had studied under him while the two were at Toronto and had joined Housman at MIT where Gusella had recently achieved his doctoral degree. Gusella was in the process of looking at opportunities to do postdoctoral work at Cal Tech under Leroy Hood, a world leader in molecular genetics, when Housman explained the opportunity to begin work on the Huntington's gene. Gusella accepted the offer in November 1979 and planned for his move to MGH. Housman told me how impressed he was by Gusella's decision to accept the risks of the new venture. There was no guarantee that success would follow nor, if it did, whether it would take years or decades.

Meanwhile, the researchers began assembling the rest of their team. Michael Conneally at the Indiana Medical Center became the group's principal population geneticist. Conneally with other workers, using the then existing protein markers, had already excluded about 20 percent of the human genome as possible sites for the Huntington's gene locus.

In March 1981 Nancy Wexler led another expedition to Lake Maracaibo, accompanied by several neurologists including Drs. Anne Young and Ira Shoulson. These workers along with many others over the years, documented the relationships within the enormous Venezuelan pedigree, determined the neurological status of the various family members, and sent blood samples on to the neurogenetics lab at MGH for DNA analysis.

By now Jim Gusella had acquired a dozen or so RFLPs, the DNA markers which he would test in his first set of experiments. Some of the markers were derived from a genetic DNA "library" that Tom Maniatis and his colleagues had the foresight to establish as a common resource for genetic investigations of the human genome in the late 1970s. One probe from the library called G8, so named after "Ginger," the technician in Housman's lab who isolated the probe, would be the third probe tested. Gusella first tested the probe on a pedigree from Indiana and achieved a lod score of 1.7, suggesting that the odds in favor of linkage of the Huntington's disease gene to the probe were 65 to 1. The lod score gives the logarithm of the odds that a genetic trait or disease links to a given marker. If the odds in favor of linkage are 10 to 1, the lod score is the logarithm of 10, which is 1.0. If the odds are 100 to 1, the lod score is 2.0. The results were extremely encouraging but not acceptable "probable proof" of linkage which, by convention, requires a lod score of at least 3.0.

By now the DNA samples and the clinical correlations from the Venezuelan pedigree were available. The results astonished the research team. The lod score had jumped to 8.4, meaning that the odds in favor of linkage between the Huntington's disease gene and the G8 probe were over 200 million to 1! A lod score that high with no "recombinations" meant that the probe had landed extremely close to the Huntington's disease gene. The probe marked a site on the short arm of chromosome 4 near the "telomere," or very tip, of the chromosome. But the gene could not yet be identified.

The discovery of Gusella, Wexler, Conneally, and their colleagues stunned the worlds of molecular genetics, neurology, and medicine. The feasibility of the ideas of Botstein's group, of Kan and Dozy's, of Solomon and Bodmer's, and of Jeffreys's was proved.

Botstein, though thrilled that his ideas had been confirmed, was still cautious and apprehensive that the early spectacular success of Gusella's group would give rise to unwarranted expectations. He feared that other research groups might not realize how difficult it would be to replicate a result like Gusella's in other diseases. Botstein realized, too, that the gene had still to be found and that an enormous amount of work lay ahead.

Even so, the Gusella group's lucky strike ultimately achieved results that in different ways justified Housman's and Botstein and White's independent objectives. There was no question that other research groups would now begin the attack on other genes. Concomitantly, the stunning initial success of the Huntington's disease project accelerated support for Botstein and White's original goal to construct DNA linkage maps over the entire human genome.

Gusella's spectacular success was not the end of his search but rather the beginning of what would be an arduously long and determined effort to identify the Huntington's disease gene. No one appreciated this more than Nancy Wexler and her colleagues, who would return to Lake Maracaibo year upon year to assess the neurological status of the native population. It was the knowledge of exactly who had and did not have the symptoms of Huntington's disease that provided the essential information that the molecular geneticists needed to hone in on the gene. No one remained closer to her patients than Nancy Wexler, who over the years developed an intimate bond with her Venezuelan family and became an eloquent champion of their humanity.

As Nancy wrote, "The most grueling aspect of the work is watching patients you have come to know and care for worsen each year. It is seeing a vibrant, intelligent, loving and beautiful little girl with limp hair, freckles, a rare but stunning smile and expressive eyes, aged 12, begin to stiffen and fall with the earliest signs of a disease, which killed her young brother and now has two other brothers in thrall. Six children of nine affected by the disease."[11]

Charles

On the day that Jeff had come to my clinic in May of 1985, it was, of course, against the background of the recent discoveries on Huntington's disease, and I realized the opportunity for similar work on the genetics of familial Alzheimer's disease. The next morning I phoned Jeff's referring physician, identified myself, and gave him the unhappy news about his patient's illness. He had suspected as much. I then asked the doctor if he had followed the recent work by Gusella and his colleagues on the linkage of Huntington's disease to a DNA marker on the short arm of chromosome 4, and I asked whether any geneticists were already working with Jeff's family in search of the abnormal gene for Alzheimer's disease. When I learned that no studies were under way, I explained that a team from Harvard, MIT, and the University of Massachusetts Medical Center had recently been established to study Alzheimer's disease. When I asked the referring physician if he would be willing to let us bring Jeff's family to Jim Gusella's attention, he readily agreed.

Even at that first moment of agreement, the physician,

who—like Jeff—was also a descendant of Hannah, expressed his deep conviction that the DNA stores we would establish on members of his family should be available in the future to other qualified investigators with worthy projects. I fully agreed. Since the diagram of the pedigree Jeff had brought was dated 1978, I asked him to update it with respect to the names, addresses, ages, and diagnoses of the family members and their spouses. I also requested copies of all previous studies done on members of the family.

Several months later the physician phoned and said that the work was now done. He would come to Massachusetts the following week, bring the records, and show us slides of the brain pathology in various affected members of the family. Our visitor was then about 65; he was an unassuming man of medium height with a warm, intelligent smile and intense, penetrating brown eyes. He bore a bit of a resemblance to Linus Pauling at the same age, his grayish-white hair was combed back, his facial expression bright and alive, his every utterance succinct and to the point.

He showed us examples of the pathology, and the findings revealed the classic and tell-tale signs of the disease just as Alois Alzheimer had described them. He described each slide with such objectivity and scientific detachment that no listener would have guessed that he was describing the findings in his own loved ones. He asked for no sympathy, he simply and directly gave us the facts we would need to know for the years ahead.

Finally, he showed a slide of brain tissue from a relative who had died of other causes relatively early on in the course of Alzheimer's disease and noted the especially early accumulation of "microvascular" amyloid in the walls of the very small blood vessels of the brain. Was this an important clue, he wondered?

He then took from his briefcase a copy of the family records, really archives, and even a floppy disk for computer display of the data. Here was all the information requested as well as data from studies that we did not then know even existed. This man was not only an expert on Alzheimer's disease, but he was clearly an expert on the storage and retrieval of pertinent medical information as well.

He provided the names of the family members on whom autopsies had established the diagnosis of Alzheimer's disease, as well as the names of the medical centers where the autopsies had been done. These records could not have been more complete or more valuable. Highly accurate medical diagnoses and, where necessary, autopsy-confirmed diagnoses are essential for a good molecular genetics study. In these "reverse genetics" studies, the scientists must work backward, starting from the clinical data-base. The geneticists must know which DNA samples come from definitely affected individuals and which come from people who, with very high probability, have escaped the disease. Moreover, not all dementias of early onset are caused by Alzheimer's disease. The twelve autopsies already done had established the diagnosis of Alzheimer's disease in every affected branch of the family and spanned three generations.

Our visitor then briefly reviewed the numerous studies in which his family had participated since the mid-1950s.[1] When he completed his review, he handed me the priceless records gathered over three decades. So it was that I first met Charles, grandson of Hannah and sole survivor of Ida's seven children. Charles had literally placed in my hands a fragment of the "Rosetta stone," a map for deciphering one of the genetic defects of familial Alzheimer's disease. In that most capricious of all lotteries, Charles had escaped her abnormal gene. With his escape and because of his and his family's enduring efforts, molecular geneticists could now begin the quest for an abnormal gene that caused Alzheimer's disease in its most classic form.

Peter St. George-Hyslop and the Link to Chromosome 21

Shortly after I received the records from Charles, my colleague, David Drachman,[1] who knew Jim Gusella personally, phoned him to see whether he had an interest in studying familial Alzheimer's disease. Jim and his recently arrived postdoctoral fellow, Peter St. George-Hyslop, then visited our medical center where we showed the records to them. Gusella was continuing to press his search for the Huntington's disease gene, which had proved to be far more elusive than was initially expected after the stunning discovery in 1983 that it was located on the short arm of chromosome 4. Heavily committed to the Huntington's work, he chose Peter Hyslop to be the principal investigator on the Alzheimer's disease project.

Hyslop was born in Nairobi, Kenya, in 1953. His parents, both English, were scientists: his father, a virologist and an immunologist; his mother, a histologist. They had arrived in Kenya to work on an important virus in veterinary medicine, but their efforts were interrupted by the Mau Mau rebellion. After living through several attacks on their field station, the parents

135

and young Peter returned to England and subsequently moved to Canada in 1968. As his education progressed, Hyslop became attracted to problems amenable to logical analysis and a reasoned scientific approach. He considered a career in one of the physical sciences or, alternatively, in the military. His parents urged him to keep his options open in the biological sciences; thus, he studied genetics and biochemistry as well as mathematics and physics while in college, and indeed finally decided on medicine as a career.

As a fourth-year medical student at the University of Ottawa in 1975, he had his first exposure to clinical neurology and examined his first patient with Alzheimer's disease. The young student was both fascinated and appalled by the disease process. Hitherto, neurology had seemed to him simply an exercise in anatomic localization of a neurological deficit either to the peripheral nervous system, the spinal cord, or a specific part of the brain in which the symptoms could be explained on the basis of an interruption to a neural pathway that served an already generally understood function. Yet in Alzheimer's disease he saw the consequences of a devastating, not yet comprehensible, process that robbed its victims of the essence of their humanity. For Hyslop, there was no greater challenge than to investigate the cause of this disease.

After graduation from medical school in 1976, Hyslop began eight years of training in internal medicine, neurology, and general methods of neurological research, but his eventual goal remained research in Alzheimer's disease. In these years, he learned as much as possible about the numerous biochemical abnormalities associated with the disease. Yet with each new paper reporting an increase or decrease of a specific neurochemical substance in Alzheimer's, Hyslop had recognized the unsolvable dilemma confronting all workers in this field: Was each new finding a primary abnormality, or was it simply one of thousands that were secondary to a still unknown initial cause? Hyslop's hard-nosed experimental bent led him to a biological science in which a first cause may be identifiable—molecular genetics.

The year 1982 was pivotal for Hyslop's subsequent career. In that year, Kay Davies, along with her colleagues in London and co-workers in Wales and Germany, linked Duchenne's muscular dystrophy to the short arm of the X chromosome using the new RFLP approach. Within months, Davies' group had discovered a second RFLP marker on the other side of the muscular dystrophy gene, thus "bracketing" the localization of this gene between two known markers. The sex-linked pattern of inheritance of Duchenne's muscular dystrophy had already shown that the disease had to reside on the X chromosome, and some suggestive evidence had been presented on roughly where the gene might be located. Hence, the discoveries of Davies's group did not have the same initial impact on the scientific community that would accompany the finding by Jim Gusella's group on Huntington's disease later in 1983. When Gusella's group used the RFLP method to link Huntington's disease to the short arm of chromosome 4, there had been no previous hint of where the gene might be located. Even so, the papers by Davies and her co-workers and a 1982 paper by David Housman and James Gusella outlining their theoretical approach to the Huntington's gene now firmly committed Peter Hyslop to a future career in molecular neurogenetics. Yet he still had to complete his residency training in neurology at the University of Toronto. During this period, he searched for clues as to where to look for an Alzheimer's disease gene.

Hyslop learned of the work of G. A. Jervis who, in 1948, had demonstrated the existence of an Alzheimer's disease-like process in the brains of many patients with Down's syndrome and the concurrent development of symptoms suggestive of Alzheimer's disease in many patients with Down's syndrome during their late thirties and forties. Several groups of investigators had subsequently confirmed and extended Jervis's work.

Also suggestive for Hyslop was work done solely with Down's syndrome. As early as 1958, a young French geneticist, Jerome Lejeune, discovered that Down's syndrome resulted from the presence of an extra copy of chromosome 21. The syndrome came to be called trisomy 21. But not until the mid-1970s did anyone find a further link betwen Down's syndrome and Alzheimer's. Leonard Heston and A. R. Mastri, working in Minneapolis, reported a small but statistically significant increase in the incidence of Down's syndrome in the relatives of patients affected by Alzheimer's disease. In 1977, based on these results and his knowledge of the increased incidence of Alzheimer's in patients with Down's syndrome, Heston suggested that a genetic defect in Alzheimer's disease might reside on chromosome 21. (The extra copy of chromosomal DNA in Down's syndrome is itself "normal": it is the fact that there are three, rather than the normal number of two, copies of a specific stretch of chromosome 21 that produces the essential features of Down's syndrome. It is usually assumed that the symptoms result from "a dosage effect"—the presence of additional, albeit normal, genes.) Thus, Hyslop had two good reasons to suspect that a susceptibility locus for Alzheimer's disease might be located on chromosome 21: (1) victims of Down's syndrome (trisomy 21) develop Alzheimer's disease with an unusually high frequency by middle age; and (2) relatives of patients with Down's syndrome have a small but statistically significant increased risk of developing Alzheimer's by late life.

Yet Hyslop knew, as did Heston, that the two associations of Alzheimer's disease with Down's syndrome did not prove that a genetic defect resided on chromosome 21. For example, not all victims of Down's syndrome developed Alzheimer's disease. Moreover, some neurologists and neuropathologists doubted that the dementia and the observed pathological changes in Down's syndrome could be taken as a valid model for Alzheimer's disease. Moreover, Hyslop recognized that the presence of a third copy of chromosome 21 might also affect the expression of a normal or an abnormal gene on some other chromosome.

In September 1983 Hyslop submitted an application to the Medical Research Council of Canada in which he proposed to study abnormalities of the enzyme phosphofructokinase in Alzheimer's disease and Down's syndrome; the enzyme was produced by an already identified gene on chromosome 21. Hyslop did not yet have any experience in the RFLP technology; thus he would use his biochemical skills to determine whether alterations in this enzyme were related to the pathogenesis of Alzheimer's disease.

Two months later, in November 1983, Gusella and his group published their classic paper linking Huntington's disease to chromosome 4. Within days of reading the findings, Hyslop wrote to Jim Gusella asking if he could come to the MGH lab to learn the RFLP techniques Gusella used so that Hyslop could apply them to Alzheimer's disease. Hyslop was thrilled when he received Gusella's reply. Jim invited him to the MGH lab and offered him the opportunity to attack Alzheimer's disease directly. (Gusella told me that at this time he already had "a twinkling of an interest" in studying familial Alzheimer's disease. One of his collaborators in the Huntington's disease work, Rick Myers, had previously collected some blood samples from a family with Alzheimer's disease which had been studied over many years by investigators at Boston University Medical School.) Jim asked Peter if he knew of any large pedigrees appropriate for the study, and Peter cited the report of a British Canadian family with familial Alzheimer's disease published by Linda Nee, Ronald Polinsky, and their colleagues at the National Institutes of Health (NIH) in 1983. Linda Nee had,

Peter St. George-Hyslop *(foreground)* with post-doctoral students Marzia Mortilla *(on left)* and Giovanna Vaula. (Courtesy of IMF Creative Communications, University of Toronto.)

in fact, phoned Gusella shortly after his group's November 1983 paper on Huntington's disease had appeared and asked him if he would carry out linkage studies on their large and extremely well-documented pedigree.

Gusella also asked Hyslop if he had any clues as to where to look for an Alzheimer's disease gene. Based on his readings of the neurological literature, Hyslop immediately suggested chromosome 21. Now the investigators had an extraordinary stroke of luck. During the course of their research on Huntington's disease, Gusella's group had established additional cell lines on the Venezuelan pedigree so that they could develop linkage maps for markers on chromosome 21, which was, of course, entirely normal in this family.

This work had been initiated by Housman's former student Paul Watkins, who was now at Integrated Genetics Incorporated. Watkins was engaged in a commercial enterprise to develop probes for chromosome 21, which could be used specifically for studying Down's syndrome as well as for locating other as yet unknown genetic defects that might reside on this chromosome. Thus, efforts to create a genetic linkage map of chromosome 21 using Nancy Wexler's Venezuelan kindred as the "reference" pedigree were well under way. Paul Watkins and his colleagues at Integrated Genetics in Framingham, Massachusetts, had already developed some single-copy chromosome 21 probes; and Gordon Stewart, first in Edinburgh, Scotland, and later at MGH, developed several others. Watkins and Stewart contributed their valuable probes, but the responsibility to discover the restriction fragment length polymorphisms, the RFLPs, which were detected by the single-copy probes, thereby making them actual DNA markers, initially fell on Watkins and Rudolph Tanzi.

Tanzi, a recent graduate of Rochester University in microbiology, had been Jim Gusella's senior research technician on the Huntington's disease project and was soon to begin his own doctoral research work at Harvard Medical School. Subsequently, Tanzi and colleagues generated the raw data for a linkage map of probes for chromosome 21. (Tanzi told me that many of the RFLPs used to construct the linkage map of chromosome 21 were found as part of his screening for RFLPs to test for possible linkage to Huntington's disease in the period before this disease was linked to chromosome 4.) Jonathan Haines, who had trained as a theoretical geneticist with Michael Conneally in Indiana, would carry out the mathematical and computer analyses required to estimate the genetic distances on chromosome 21 between the fragments detected by the probes.

In the early fall of 1984, Rudolph Tanzi's task to complete a map of chromosome 21 suddenly took on new urgency. A just published paper in the August 16 issue of *Biochemical and Biophysical Research Communications* by George Glenner and Caine Wong revealed a thrilling breakthrough. These authors had isolated and purified the amyloid protein that abnormally accumulates in the walls of the small cerebral blood vessels and had gone on to sequence the first 24 amino acids. Glenner and Wong discovered that the amino acid sequence in the vascular amyloid of Down's syndrome was very similar to the cerebrovascular protein found in Alzheimer's disease. They believed their discovery to represent "the first chemical evidence of a relationship between Down's syndrome and Alzheimer's disease."

Glenner and Wong ended their paper with a bold prediction: "Assuming β-protein is a human gene product, the presence of a common amyloid protein in both Down's syndrome (trisomy 21) and Alzheimer's disease suggests the possibility that the genetic defect in Alzheimer's disease (whether acquired or inherited) is localized to chromosome 21."[2] Tanzi brought the article to Gusella's attention. Gusella now set the division of labor between Hyslop—soon to arrive—and Tanzi. As they had previously agreed, Hyslop would carry out the genetic linkage work on Alzheimer's disease, while Tanzi would complete his linkage map of chromosome 21 and go after a possible gene for amyloid.

When Peter Hyslop arrived at MGH in January 1985, another bit of luck awaited him. A substantial number of blood samples from the large British Canadian family described by Linda Nee and Ronald Polinsky and their colleagues in 1983 had already arrived at the MGH lab. In fact, Tanzi had already tested these DNA samples for possible linkage to the then available chromosome 21 markers but found no significant linkage.

Linda Nee, a social worker at the National Institutes of Health in Bethesda, had discovered this large family quite by accident in 1978, when a letter arrived from a Canadian physician inquiring about what could be done to investigate the plight of the wife of one of his colleagues and their family. Nee wrote back, asking the physician to ascertain whether the affected members would be willing to come to the clinical center at NIH for diagnosis. The diagnosis of familial Alzheimer's disease was readily established, and the NIH group went about documenting the line of transmission of the disease in previous generations.

The first known cases of Alzheimer's in this family were traced to the children of a woman who was born in Northumberland, England, in 1763, not far from the coast of the North Sea. This woman and a number of her children immigrated to the New Brunswick area of Canada beginning in 1837. The mother lived into her nineties and thus certainly did not carry the Alzheimer's gene; her husband who died young was presumed to be the carrier. In time, numerous descendants of the couple developed dementia in middle life, although the earlier generations did not recognize the cause. By the 1960s and 1970s, descendants of the couple, now in the seventh generation, were falling victim to a mysterious dementia generally apparent by their early or mid-fifties. Eventually, the NIH group evaluated 51 members of the family.[3] The symptoms of the affected members were classic for familial Alzheimer's disease, as were the neuropathological findings in several of the cases that subsequently came to autopsy. The NIH workers realized that the pattern of inheritance was consistent with that of an autosomal dominant gene with complete penetrance, just as Charles had previously established in his own family. (An autosome refers to one of the 22 pairs of nonsex-determining chromosomes; dominant refers to a pattern of inheritance in which one-half of at-risk individuals inherit a gene; and complete penetrance describes that type of genetic expression in which everyone who inherits the defective gene eventually develops the disease.)

As I read the report from the NIH group, I was struck by the fact that the characteristics of the disease in this family of British Protestant origin were virtually identical to those in the Byelorussian Jewish family with whom I was working. The clinical symptoms in the two groups of affected individuals were indistinguishable; the neuropathological findings were essentially identical; even the mean age of onset of the disease in the two families was virtually identical. Yet the number of living affected members in each of the two families from whom DNA could be obtained was still relatively small for studies of genetic linkage analysis. In Charles's family, there were only four living affected patients from whom blood samples for lymphocyte cultures could be obtained, whereas in the British Canadian family, there were only five such individuals, even though there had been 54 documented cases of the disease over the past eight generations. In the interim, the MGH group had also acquired a small but extremely well-documented multigenerational family living in Germany, followed there by Dr. Peter Frommelt, which had the classic characteristics of Alzheimer's disease, including a mean age of onset very close to that found in the other two families.

The subsequent course of the genetic linkage studies would be profoundly affected by yet another chance event. In May 1985, at about the time Jeff came to my clinic, Rick Myers, the genetic epidemiologist who was pursuing research on Huntington's disease and carrying out genetic counseling for this disorder, wandered into Hyslop's lab and asked if a neurologist could come with him to draw blood from a patient who might have Huntington's disease. Since the patient's symptoms were minimal, Rick also wanted the neurologist to provide a "second opinion."

Hyslop remembers that as they drove off in Myers's gold-colored old Dodge toward the patient's home in Newton, Myers asked Hyslop what he was working on at MGH. Hyslop described his interest in the genetics of familial Alzheimer's disease and related his attempts to acquire informative pedigrees. At the next traffic light, Hyslop continued, "You know, there are these two sets of specimens in Jim Gusella's freezer numbered 1740–1748 and 2516–2526 labeled '? AD'." Hyslop went on to tell Myers that neither Jim Gusella nor Wendy Hobbs, one of Jim's chief technicians, had the pedigree diagrams but that both sets of specimens bore surnames that were Italian.

Stopping at the next traffic light, Myers astonished Hyslop, "I collected one set, and I can give you the pedigree diagram." Myers had collected the samples from a large Italian pedigree that Robert Feldman, now chief of neurology at Boston University, had been following since the early 1960s, when he and colleagues then at Yale Medical Center first described this pedigree. Several years earlier, Myers had assumed that the specimens would eventually become invaluable for linkage studies of Alzheimer's disease, but now his main interest had become Huntington's disease. Thus, he had not yet pursued the work on the Italian family. Meanwhile, Feldman had given all the information on the Italian pedigree to Ronald Polinsky at NIH, who had agreed to follow the family from then on for purposes of further genetic study. Moreover, Polinsky had made a trip to Italy and obtained blood samples from a family with similar symptoms. Myers suspected that the two sets of samples were from different branches of the same enormous Italian pedigree that was now being intensively studied by scientists in both France and Italy. A week later, Myers had dug out pedigree diagrams and confirmed the link.

The scientific study of this Italian kindred goes back to 1963 when Feldman and his colleagues described a family afflicted with dementia and several other striking symptoms. In addition to their mental deterioration and aphasia, most of the patients also suffered from seizures or had brain wave (EEG) abnormalities suggestive of a seizure disorder. In the later phases of their disease, their faces remained immobile, their limbs were held as if in a "plastic rigidity," and their body postures, even while lying in bed, were contorted, with their upper and lower limbs flexed in an exaggerated manner toward their body. There had been 13 cases of the disease over the previous four generations. Signs of forgetfulness appeared extremely early, between the ages of 30 and 40, and in 8 of the 13 patients the disease began in their thirties. Disease onsets as early as age 31 and age 32 occurred in two cases.

When Feldman's group examined the temporal lobes of these patients, they found numerous senile plaques, neurofibrillary tangles, and granulovacuolar changes fitting the diagnostic criteria for Alzheimer's disease. They noted, too, that neurofibrillary tangles were found in parts of the basal ganglia, such as the claustrum, putamen, and globus pallidus, where the pathology might explain the disorders of movement and posture, and in the dentate nucleus of the cerebellum, where the pathology might explain the spontaneous jerky contractions of individual muscle groups which were also very frequent in affected members of this family. The researchers even noted degeneration of the anterior horn cells or "motor neurons" in the spinal cord, which they presumed had led to the atrophy of some of the muscle groups.

Feldman's group emphasized that "the constellation of neurologic abnormalities in these patients indicates that their condition was more than just dementia."[4] They emphasized that the condition described affected neurons of all types in a widespread distribution. Although the findings in the cerebral cortex were classic for the diagnosis of Alzheimer's disease, much more was involved—namely, the especially early age of onset, the seizures, which often occurred very early in the course of the disease, and the disorders of movement and posture found much earlier in the course of the disease than in the cases described by Alois Alzheimer and his contemporaries.

Even so, Feldman's group realized the great research importance of this family and went as far as they could given the techniques then available (in the 1960s). Careful electron-micrographic studies were done in 1965. These findings of the cerebral cortex were also consistent with the diagnosis of Alzheimer's disease. In addition, the research group attempted genetic linkage studies with the markers then available, but no relationships could be found.

And so the matter rested for many years. In 1973 Jean-François Foncin and V. Supino-Viterbo discovered a family in France with characteristics identical to the family described by Feldman's group. Foncin's discussion with the spouse of a patient revealed that the family had originated in the South of Italy. Wasting no time, Foncin followed up his lead.

Foncin discovered that there was an extensive branch of the family in Italy as well. Amalia Bruni and her colleagues in Italy investigated the Italian branch, while Foncin continued to study the French patients. In 1985 Foncin, Bruni, Feldman, and six other collaborators published an extraordinary paper in which they established that the American, French, and Italian victims were all descended from a common ancestor.[5] They documented 43 cases of the family disease among 1,435 descendants of the common ancestor over the past ten generations. They also proved that environment played essentially no part in the manifestations of the disease since the affliction "touches in the same manner the American, Italian and French branches of the family."

By 1988 Foncin, Denise Salmon, and Bruni had traced the family back even further. There were now 4,000 known descendants, with 60 known cases of the family disease, of whom 5 were still living. All were descended from a common ancestor, a woman called Vittoria, born in 1715 in a small mountain village in the provence of Calabria at the southern tip of Italy. With the blood samples previously obtained by Rick Myers and Ron Polinsky's group and with new samples obtained by the French and Italian workers, the MGH group now had blood samples on seven affected individuals and many more samples from assured "escapees" from this large pedigree.

As Hyslop began his work, he planned to add together the lod scores from the four families, assuming that the abnormal gene in the four kindreds might be the same. Though aware that the gene in the Italian family might be a different gene from that in the other families, Hyslop tentatively assumed that he was seeking a common gene, even though the specific mutation causing different forms of the disease in the various families might, indeed, be quite different.

There was already much precedent for Hyslop's tentative assumption. By the early 1980s both Duchenne's muscular dystrophy and a milder but similar muscle disease called Becker's dystrophy were generally thought to be caused by different mutations of the same gene. Confirmation came in 1983 when Kay Davies and her group suggested that both conditions were linked to the same general region of the X chromosome and thus were, in fact, the result of different mutations of a common gene. By late 1985, as Hyslop began his studies, all known cases of Huntington's disease in different families had been linked to the same region of chromosome 4. As the MGH group's work on Alzheimer's disease progressed, the actual gene causing Duchenne's muscular dystrophy was isolated by a large international collaborative group led by Louis Kunkel at the Children's Medical Center in Boston. This gene, about 3 million base-pairs long, was thus far the largest gene discovered. Different mutations within this enormous gene produced diseases of profoundly varying severity.

Thus, if Hyslop's hunch was correct, the disease in the four families should link to a common locus. The MGH group initially tested markers in the region of chromosome 21 that were associated with the essential region responsible for the main features of Down's syndrome. The markers for this region gave negative results.

They then tested a series of markers "above" or toward the centromere (the region joining the long and short arm of the chromosome), away from the region that must be "trisomic" for full expression of the Down's syndrome phenotype. Now two of the markers tested—identified according to the convention of the geneticist as D21S16 and D21S1/D21S11—gave positive lod scores; the markers D21S1 and D21S11 were so close together that they could be considered as testing a common locus. The D21S16 probe gave a substantial positive score of 2.56 in the Italian pedigree. Thus, the odds of linking their disease to this probe were about 300 to 1. However, the other pedigrees showed essentially no linkage to this marker. The second set of markers—D21S1/S11—gave positive lod scores in three of the pedigrees for a total lod score of 2.35. (The score in the German pedigree was very low; this pedigree was minimally "informative" for the second set of markers because the same form of each marker—that is, the length of the restriction fragment—tended to be present in both parents of affected members of this pedigree.) Given that two relatively closely spaced markers gave substantial positive scores for the four kindreds taken together, the statisticians carried out a "three-point linkage analysis" (one "point" is for the presence of the disease, the other two for the markers) and estimated the odds that the disease was linked to the general chromosomal region spanned by the two markers. They then obtained a peak lod score of 4.25, giving strong odds that the disease was linked to a region of the chromosome near these markers.

The paper by our group, titled "Genetic Disease Causing Familial Alzheimer's Disease Maps on Chromosome 21," was published in *Science* in February 1987.[6] Commentaries in the journals *Nature* and *Science* appraised the significance of the achievement. Yet the lod scores in favor of linkage in Charles's family and in the British Canadian family were each only about 0.61, which meant that the odds in favor of linkage in each of these families taken separately were only 4 to 1. The statistical evidence in favor of linkage was obtained only by adding together the results from all four pedigrees—the British Canadian, the German, the Italian, and Charles's family—where the results of the Italian pedigrees dominated the total. Yet, if the same gene was indeed involved in all four families, then the evidence was strong for linkage of the disease to chromosome 21 in the several pedigrees.

Even as our paper was published, a number of other groups made another very important discovery, which further pointed to a probable role of chromosome 21. Four groups independently discovered that a gene encoding the precursor of the amyloid found in the senile plaques of Down's syndrome, Alzheimer's disease, and normal aging was located on chromosome 21.

Excitement ran high. Could the genetic defect in Alzheimer's disease in fact be an abnormality in the gene that produced the amyloid precursor protein? Was the localization of both genes on chromosome 21 of great significance or merely coincidental? A number of research groups worldwide now turned their attention to a gene on chromosome 21 that encoded the amyloid precursor protein.

A Gene for Amyloid

In the early 1980s, an intensive search began for the structure of the "peculiar substance" that Alois Alzheimer and his contemporaries had found within the small blobs of degenerating axonal fibers and brain cells collectively called "senile plaques." Paul Divry had called this substance "amyloid" in 1927. By 1983, M. Kidd (the same researcher who had carried out electron-micrographic studies on amyloid in the early 1960s) and his colleagues had isolated and determined the amino acid composition of the amyloid protein found in the core of the senile plaques.

Within a year, George Glenner and Caine Wong had purified the amyloid protein often found in the walls of the very fine blood vessels of patients with Alzheimer's disease. They then determined the sequence of the first 24 amino acids of the amyloid fragment. One of their objectives was to determine whether this protein was unique for Alzheimer's disease and the Alzheimer's disease-like state that so often accompanies Down's syndrome, so that a diagnostic test for Alzheimer's disease could be devised. If the amyloid protein were really unique to

Alzheimer's disease, then they reasoned that they might develop a diagnostic test in which an antibody would recognize an amyloid-related protein in the blood of patients with Alzheimer's disease. This objective was not realized, partly because the accumulation of amyloid in the small cerebral vessels is not specific to Alzheimer's disease but can occur in normal aging as well. Moreover, the postulated blood proteins had not been found in more than trace amounts with any consistency.

Yet Glenner and Wong's results would soon lead to an even more important and new direction. Their second germinal paper in 1984 posed the question of whether the closely related amyloid proteins that they had found in cerebral blood vessels in both Alzheimer's disease and Down's syndrome could be the product of a human gene. Within a year, Wong and Glenner further discovered that the amyloid found in the senile plaques within the brain was quite similar in structure to the amyloid found in the arteries of the brain.

Concurrently, Colin Masters then at Perth, Western Australia, and Konrad Beyreuther and his colleagues largely from the Institute of Genetics at the University of Cologne went on to purify the amyloid protein found in the plaques in Alzheimer's disease and Down's syndrome. They discovered that the protein consisted of multiple aggregates of a polypeptide that in some cases consisted of about 40 amino acid residues. The amyloid fragment had essentially the same structure whether found in the brains of patients with Alzheimer's disease, Down's syndrome, or normal elderly subjects.

Beyreuther's group reasoned that the β-amyloid protein, which they called A4, was actually a small fragment or cleavage product of a much larger precursor protein. Now with the precise sequence of the A4 subunits determined, researchers could immediately predict the sequence of the coding nucleotides of DNA and RNA that would direct the synthesis of just such a protein. After synthesizing corresponding probes, the researchers could work backward and attempt to find the chromosome and then the gene which had a base sequence that matched the sequence required to predict the amino acid structure of the A4 amyloid.

Because coding regions of genes (the exons) are broken up by intervening noncoding sequences called introns, the investigators could not assume that the coding sequences of nucleotides along the amyloid precursor gene were connected directly to each other. Thus, they had to construct a series of smaller probes, called

Dr. Colin Masters. (Courtesy of Dr. Masters.) Dr. Conrad Beyreuther. (Courtesy of Dr. Beyreuther.)

oligonucleotide probes, which had a greater chance of finding and then binding to a corresponding section of the sought-after gene. Publication of Glenner and Wong's work on microvascular amyloid and Masters and Beyreuther and their colleague's findings on "plaque" amyloid opened the way to the search for a gene encoding an amyloid precursor protein, a protein which when cleaved by some yet unknown reaction would yield the amyloid fragments found in the fine cerebral blood vessels and senile plaques of Alzheimer's disease, Down's syndrome, and normal aging.

Groups around the world now raced to find the gene encoding the amyloid precursor protein. Remarkably, within a single month—February 1987—four different groups had discovered the gene's location: all agreed that it was localized on chromosome 21. Two such groups reported their discovery in the very same issue of *Science* in which our group (Peter St. George-Hyslop and colleagues) reported the linkage of familial Alzheimer's disease to chromosome 21. One of these two groups was led by Dimitri Goldgaber, Carlton Gadjusek, and their colleagues at NIH.[1] (They had actually announced their discovery at a neuroscience meeting in the late fall of 1986, but publication would take several months.)

The second of these groups included Rudolph Tanzi, who was still a graduate student working under Rachel Neve in the laboratory of David Kurnit and colleagues.[2] Now as a consequence of his long-standing interest in creating a linkage map of chromosome 21 and of the Hyslop group's concurrent report that familial Alzheimer's disease was linked to chromosome 21, Tanzi and his associates had the unique opportunity to estimate the approximate distance between the amyloid gene and a putative gene for familial Alzheimer's. They placed the two genes roughly within the same region of chromosome 21, and raised the intriguing possibility that the two genes might be identical.

In the very same week that these papers came out in *Science* in the United States, the English publication *Nature* carried a detailed characterization of the amino acid sequence of the amyloid precursor protein and localized its gene to chromosome 21. This impressive effort began at the University of Cologne when Benno Müller-Hill, a bacterial geneticist, set his student, Jie Kang, on the project. She isolated the c-DNA clone which was then quickly sequenced. The results established that the amyloid precursor protein contained 695 amino acid residues.[3] Based on their understanding of the predicted protein structure, Kang and her colleagues realized that the amyloid precursor protein had features characteristic of a specific type of cell-surface receptor. Receptors are specialized structures on the outer surface of a cell's membrane that interact with a chemical messenger from another cell; the interaction sets off a cascade of biochemical changes within the recipient cell. The workers went on to propose that the sequence of the amyloid precursor protein, together with its localization on chromosome 21, strongly suggested that the cerebral amyloid deposited in Alzheimer's disease and Down's syndrome might be caused by the aberrant breakdown of a cell-surface receptor.

Also in February 1987, another research group including Nikolaos Robakis, Henryk Wisniewski, and their colleagues at the Institute for Basic Research in Developmental Disabilities in Staten Island, New York, reported in *Lancet* that the amyloid gene was on chromosome 21.[4] They, as well as the other three groups at work on the amyloid gene, suggested that the symptoms of the Alzheimer's disease-like pathology found in Down's syndrome might be due to a single highly specific "dosage effect"—the result of the presence of three genes rather than the normal two that code for the amyloid precursor protein.

Hopes were now high, though cautious, that the cause of

Alzheimer's disease might become known within months. Within three weeks, Jean-Maurice Delabar along with Dimitri Goldgaber, Paul Brown, Carlton Gadjusek, and their colleagues at NIH reported that they had discovered an extra copy of the amyloid gene in patients with Down's syndrome. While this finding was expected because of the presence of an extra chromosome 21 in this condition, they also claimed to have found an extra copy of the amyloid gene in three patients with sporadic Alzheimer's disease. If they were correct, the puzzling cause of Alzheimer's disease was essentially solved: either an extra copy of the amyloid gene or an abnormal amyloid gene might produce the symptoms of Alzheimer's disease. The efforts now concentrated in two questions: First, would the findings of Delabar's group be replicated by other workers? Second, was the amyloid gene the culprit causing Alzheimer's disease?

But the hopes of the spring began to evaporate as quickly as they had arisen. By September, two groups had discovered polymorphisms or RFLPs for noncoding regions of the amyloid gene. Thus, they were able to test directly whether a given variation linked tightly to the presence of Alzheimer's disease in specific families. In the September 10, 1987 issue of *Nature*, Rudolph Tanzi, Peter St. George-Hyslop, and colleagues showed that familial Alzheimer's disease in all four of the original families studied was not tightly linked to the gene for the amyloid precursor protein. In the same issue, Christine Van Broeckhoven, together with numerous distinguished European and Australian researchers, reported the same result.

Then, only a month and a half later, in the October 30, 1987 issue of *Science*, our large research group headed by Peter St. George-Hyslop reported an absence of duplication of chromosome 21 genes in both familial and sporadic Alzheimer's disease, thus failing to confirm the earlier report of Delabar and his colleagues. Now it appeared possible that the appearance of amyloid in the brain of patients with Alzheimer's disease—at least in the patients studied thus far—might merely be a by-product of neuronal death rather than a primary cause of the disease. Even so, the possibility that the amyloid protein played an active but secondary role in the pathogenesis of the disorder remained open. Two other papers in the same issue of *Science*—one from the Rudolph Tanzi-Rachel Neve group and the other from Dennis Selkoe's group—also reported the absence of a duplicate gene that encoded the amyloid precursor protein in Alzheimer's disease.

It now appeared that identifying the abnormal gene(s) that caused familial Alzheimer's disease would be accomplished neither quickly nor easily. Perhaps, as some speculated, an abnormal gene somewhere else on chromosome 21 might modify the expression of an otherwise normal β-amyloid gene. The role of β-amyloid itself in the pathogenesis of Alzheimer's disease remained open to question. We were now asking the same questions that Alois Alzheimer and his colleagues had asked some eight decades ago, but we were now asking them at the molecular level. Even so, the answers were still unclear and unsatisfactory. The relative closeness, perhaps 8 to 10 million base-pairs, between the amyloid gene and the putative Alzheimer's disease gene might be sheer coincidence. The distances, though relatively close in terms of the length of the entire genome, were still rather large for the two genes to interact directly.

When the gene for amyloid was discovered, we had a "candidate gene" that could rapidly be tested using existing techniques. Now with our candidate gene shot down from contention, at least for the families studied to date, the research groups had to return to the linkage method of attempting to get ever closer to a putative gene. That meant that the geneticist would still have to rely on the family method and the continuing documentation of affected individuals and probable escapees in the families that we were now studying, with an ongoing search for additional informative families. The cautious hopes of an early solution to the cause of Alzheimer's disease in the spring of 1987 might not have outlasted the fall, but the final objectives remained as important as ever, and the work would go forward.

A Siege of the Soul

Amid all the research advances on the molecular genetics of Alzheimer's disease that lent excitement to my scientific life, I now faced a different kind of struggle against the disease with respect to my personal life. Several months after Jeff came to my clinic, the lives of my own parents began to unravel. My father, David, an 86-year-old retired physician, had always prided himself on his self-reliance and independence. The only concession he had made to aging had been a move to Florida where he could escape the New England winters that constantly threatened to flare up the asthmatic bronchitis he had developed after a pneumonia in the years before antibiotics were available. But that summer he grew weak and severely anemic. In the early fall he was diagnosed as having chronic myelogenous leukemia, a disease unusual at this age except in those with increased exposure to radiation, which my father, as a radiologist, had had. With treatment, the mean life expectancy was four years, and for several months my father responded reasonably well. That same fall my mother, Anne, then 78, began to show a very rapid progression of a dementia that had appeared abruptly in her early seventies after her physician had added a new antihypertensive medication to her regimen. Miraculously, after some changes in therapy had been made, the illness had seemed to be in remission for the next three years. But by age 75, the signs of progressive disease became clear and unrelenting.

We had always prized our mother for her kindness, emotional warmth, and intellectual ability. She effortlessly whizzed through college and law school in a combined five-year program, finishing at age 21 and graduating second in her class. She then left her native West Virginia, arrived in New York City, and quickly passed the New York bar. By now the Depression of 1929 had all but destroyed the job market, and as a woman, she was unable to find a position as a lawyer with a major firm. Instead, she found work as a law librarian. In 1933 she met my father who was then interning in New York; after their marriage they returned to his home ground in Massachusetts. My mother never returned to the law, but later, because of her love of history and literature and her need to contribute financially to the costs of a college education for me and my brothers, she became a teacher of English literature for eleventh- and twelfth-grade students. This love of literature had remained even into these declining years.

By early December 1985, my father's remission had ended. He was now in the final stages of acute myelogenous leukemia, with only weeks to live. In those few months, my mother's decline was equally terrible to witness. Although we may often mentally rehearse what the death of a parent will be like, especially that of an aged parent, it is said that no one can really prepare us for it. Nonetheless, in time the wounds do heal, and we grip the death of another as a fact of life. However, the circumstances of my mother's illness immersed us all in a hell that we had never contemplated, let alone rehearsed.

Almost weekly my mother's decline was evident. She, who only a decade before could probe the salient points of contemporary issues in constitutional law with expert precision and crisply recall the plots of English classics read years ago, could now not even find her way from room to room. By March, admitting her to a nursing home—a thought that had been unspeakable just a few months earlier—became a reluctant choice. As a neurologist as well as a son, I knew that little more could be done, but that could not be my decision alone. My mother was admitted to the care of another neurologist who confirmed that her dementia was indeed untreatable.

The agonizing search for an acceptable nursing home began. Fortunately, there was soon a vacancy, and a single room at that, in a home in our own town only about three miles from our address. My wife Linda and I tried as best we could to make the room look homey, placing familiar photographs on the walls and fond mementos about the room. When the weather was good, we often brought my mother home for a Sunday dinner or in the early days to a restaurant nearby. But her decline continued unabated, and it was agonizing to watch. The visits to the nursing home became like a funeral every weekend, except for one saving grace. My mother and my youngest son, Alex, then four and a half, had already formed an affectionate bond. He frequently accompanied me on the visits to the home; this golden-haired, bright-eyed child brought youth and hope to the visits in unexpected and offhanded ways. His visits briefly transformed a mere existence into living for my mother as well as for many other residents of the home.

In those difficult days I drew what inspiration I could from the courageous examples that Charles and his family had set. Although my anguish over my mother's decline continued to tear at my heart, I was also acutely aware that I could not let my personal considerations impair my objectivity in my work on Alzheimer's disease. Charles and Ben had lived and worked through almost infinitely worse torments; yet they retained their objectivity even as their commitment to fighting Alzheimer's grew. And so would I.

By the late fall of 1986 my mother's progressive Parkinsonian-like rigidity of limbs, which often accompanies the later stages of some dementias, had worsened. This symptom had been successfully treated with low doses of Sinemet for several years but was now dramatically worsening. I asked her physician at the home to consider raising the dose still further, and, if this did not work, to add another agent, often useful in such conditions, called bromocriptine. This drug could reduce the rigidity, though at great risk of creating ever more confusion in an elderly, already confused patient. Before her physician could implement my suggestions, her rigidity worsened. I then received a call from the nursing home advising me that the staff recommended removing my mother from the Level III facility to the Level II unit inasmuch as she now required more nursing care as a consequence of the severe rigidity. I asked if they would wait until the increased doses of Sinemet and the Parlodel could be tried, but the director of the home insisted that I come and see the Level II facilities.

In Level III there is some semblance of normal life. There are communal dining areas, lounges, television sets, and a camaraderie between nurses and patients. As a neurologist I, of course, knew the term "Level II" but had never personally experienced its sights and sounds; the moans, shrieks, and sighs alone would preclude my ever bringing Alex to visit his grandmother. I shuddered at the thought that my mother might have to die on Level II. I could not fault the nurses, physicians, or nursing home administration for conditions in Level II; these simply reflected the advanced stages of the patients's own incurable and untreatable diseases.

Happily, the bromocriptine worked and my mother remained on Level III. By mid-winter 1987, she was even walking with assistance once again. By the middle of March, the snows had melted, and one Saturday I eagerly told her that the next day would be warm enough for me to bring her to my home once again. But on that Sunday, March 15, 1987, I received a call that my mother had died suddenly of a heart attack. I thought for a moment that I would have liked to have brought her home that one last time, but I was comforted by the thought that she had finally obtained the release she had wanted for several years. The siege of her soul had ended. Death had brought her the only surcease yet known from Alzheimer's disease.

Elizabeth W. and the Second Family from Bobruisk

By the late spring of 1987, not long after my mother's death, it was already clear that if we were to accelerate the pace of the genetic linkage work, we would need to find more large kindreds with familial Alzheimer's disease. From the medical literature I knew of one kindred ideal for such studies. The 1981 paper by Goudsmit, Gadjusek, and colleagues on familial Alzheimer's disease had been based on the study of two families. The disease in one family, that of Hannah's descendants, had possibly originated in Bobruisk; the first known case in their second large kindred had assuredly been born in Bobruisk. By 1987 Gadjusek's group was studying the possible role of the amyloid gene in Alzheimer's disease and was not directly carrying out linkage analysis. I therefore asked Peter Hyslop if I should try to obtain the participation of this second kindred in the genetic linkage studies. Succinct as always, he simply responded, "Absolutely." Obviously, the help of this family would be of great significance. If the genetic defect in this family was identical to that in Hannah's family, then the lod scores could be pooled with confidence.

This other family, the "M.P. family," had been investigated in almost as much detail as Hannah's. Robert Terry and Robert Katzman had studied this family earlier, and, as in the case of Hannah's descendants, tissue from the brains of affected famly members had been injected into the brains of primates that Dr. Gadjusek and his co-workers studied, eventually to prove that familial Alzheimer's disease was not transmissible. Moreover, there were six documented autopsies of family members with proved Alzheimer's disease.

After several phone calls, I obtained the name and phone number of the family member, Dr. P., who 25 years ago had initiated his family's participation in these research studies. But by now he was shattered by the ever-present spectre of Alzheimer's in his family. He told me how the family had been "decimated" and that every time he attended a wedding or family gathering the conversation always ended up on which member might now be developing the disease. He appreciated our efforts, but he had had enough; it was too much to ask. He regretted the decision but he could not ask his family to participate. He agreed, however, that I could write him and outline the present status of the work and what we would want the family members to do. I did this in a letter of June 16, 1987 and added,

> I can understand the emotional turmoil that you must be going through in trying to decide whether or not you wish for your family members to participate in this research. What I can tell you quite honestly is that there is now hope for an understanding of this disease process and in time for rational therapy. I cannot predict when this time will come but given the recent revolutions in molecular genetics there is much reason for hope. The patients and families who have worked with us share in that hope and I believe that their commitment to this endeavor has added to the quality of their lives by reason of such hope. We now have a chance to strike back against Alzheimer's disease.

After he received the letter, we spoke again. While he understood and accepted our motives, he still felt that the strain on his family would be too great. So the matter ended, or so I thought. On July 1, 1987 I received a telephone message to call a Mrs. Elizabeth W., a cousin of Dr. P. Perhaps this was another chance. I mentally rehearsed the answers to all the questions I could anticipate and apprehensively picked up the phone to call Mrs. W. But Mrs. W. asked no questions. Instead, a committed voice at the other end of the line simply said, "I have read your letter. We are in. You can call me anytime of the day or night." It was only several months after my own mother's death, and her resolve struck a resonant chord. Her anguish, as had mine, had been transmuted to action. I knew we could count on Elizabeth W.

Elizabeth herself was now 65. Both great-grandmother and grandmother had died of Alzheimer's disease; her at-risk father, though he died too young to have symptoms, had apparently transmitted the disease to her brother and a sister, both of whom had now died. Elizabeth was only 15 when her father died, and she assumed responsibility for the house so that her mother could work outside the home. Later, Elizabeth was left to care for her sister's four children of whom one, then in his early forties, had developed the disease and had by age 45 entered a nursing home.

But Elizabeth, recently widowed after 43 years of marriage, was also caring for her elderly 88-year-old mother who suffered multiple medical problems. Despite all this responsibility, Elizabeth had the strength to work with the local chapter of the ADRDA— Alzheimer's Disease and Related Disorders Association—and coordinated her family's efforts to provide us with a complete pedigree and current addresses of family members. Whenever necessary it was Elizabeth, herself directly or indirectly through appropriate family contacts, who would convince a reluctant family member to join the study. Elizabeth's decision to bring her family into the study and her tireless efforts would in time make a valuable contribution to the quest for abnormal genes.

I once asked Elizabeth how she had found the strength to cope with so much adversity. "You know, Doctor, no one ever asked me that, and I have never really thought about it. I never thought I'd done anything remarkable. Because I was the healthy one, I just helped the others. It seemed so natural."

Once again, I had drawn renewed vigor from the example of the families of my patients. And so had Dr. P., who eventually rejoined the fight that he had once begun.

A Unity of Heart and Mind

By late 1987, even after Elizabeth's family had joined our study, our progress toward finding the abnormal gene was still limited both by the relatively small number of families available for study and the small number of affected individuals within each family. Our situation was thus very different from the experience of researchers of Huntington's disease, where over 100 affected individuals from within a single family had been identified on the shores of Lake Maracaibo. Unfortunately, no such family existed for us. And so it became all the more necessary for us to discover additional families affected by Alzheimer's disease and to screen all at-risk members of the large families we were presently following.

At this time, over 50 families with familial Alzheimer's disease were already well described in the world's literature. However, most of these families were rather small with only a few, if any, living affected individuals. Even so, members of virtually all these families had already been contacted by one or another of a community of scientists at work on Alzheimer's research around the world.

From our own clinic population we discovered a small family of French Canadian origin with two living affected cases, the minimum number that would be of value to us. New families were discovered in England and even several small ones from Northumberland, from where the large British Canadian family had originated, but none of these could be traced directly to the first known affected ancestor of the original family studied.

Japanese scientists had ample numbers of patients with late-onset Alzheimer's disease to study. However, very few Japanese families with early-onset familial disease were then known. A small family of Japanese ancestry was found living on a Hawaiian island and was now cooperating with a Seattle team of researchers.

Families as large and as well documented as the two families of Byelorussian origin, the British Canadian family, and the Italian family from Calabria probably still numbered fewer than ten in the world. The larger the family, the greater the opportunity for useful results. For in these families the suspected bit of abnormal DNA must be identical to that found in every other affected case and differ significantly from the comparable strip of DNA in every unaffected family member.

Therefore, it was necessary for us to screen all individuals at risk within the age of risk in all the families we were following, especially those in Charles's and Elizabeth's families. Between these two families there were 34 individuals at 50 percent risk and a number of others whose at-risk parent had died before reaching the age at risk and who were thus at 25 percent risk.

The ethical considerations involved in deciding to screen these individuals were of great concern to us. People, many of whom had never heard of us, would have to receive a letter or a phone call reminding them of their risk for familial Alzheimer's disease, that we knew about it, and that we wished to determine their current status with respect to the presence or absence of symptoms. We could have waited years if necessary, until the advent of the disease took its toll and left patients who needed diagnosis and whatever treatment we could give, but that would be years lost in attempting to find the gene so that we could get to the source of the problem. We decided to contact everyone at risk and explain the situation candidly to them. Most, we knew, were free of symptoms now, but some might develop the disease in future years. Even some of those who had escaped the disease would likely go through periods of anxiety and depression as they entered the age of risk. Our baseline studies might in time offer something positive to these people, aside from helping us find the abnormal gene in their family sooner.

Any individuals who felt they already had a memory problem were invited to come to our medical center for a complete neuro-logical and neuropsychological evaluation. I would see each patient, and David Drachman would provide a second independent opinion when he was available. At this time, David was the chair-man of the Medical and Scientific Advisory Board of ADRDA. Dr. Brian O'Donnell, our chief neuropsychologist, and his associate, Joan Goodwin Swearer, a doctoral candidate, carried out the neu-ropsychological studies. Their evaluations were done indepen-dently of mine and David's. Only if all examiners agreed that a patient had Alzheimer's disease or had clearly escaped the disease would a decision be sent along to Peter for entry into his calcula-tions of lod scores.

At one level, we were convinced that we had to find new affected cases so that the molecular genetic studies could progress. (It was equally important for us to identify those individuals who had reached the above 60-year-old age group and had happily escaped the disease with very high probability.) At another level we dreaded having to diagnose Alzheimer's disease in anyone, especially once we had met the person and his family. It was espe-cially painful to diagnose the disease in young adults with early symptoms. Some of the people we studied demanded to know the results of our investigations. Others insisted that we tell them nothing.

As we began the screening, Peter called one day and told us he had discovered the whereabouts of Hannah K. who had been living at a nursing home for the past seven years. He asked that I check on her status. Hannah's daughter agreed that I could visit her mother. Hannah, now in her mid-seventies, lay quietly in a clean bed in a modern nursing home, but had not spoken for the past three or four years and had to be spoon fed. I was looking on the last living affected grandchild of Hannah. This Hannah, an eldest daughter, had been given her name in memory of her deceased grandmother. The name had not been a lucky one in this family.

My examination revealed no "focal findings" that might have been suggestive of a stroke or other intracranial catastrophe, but it was consistent with, yet not proof, that she had Alzheimer's dis-ease. A computerized axial tomographic or CAT scan of the brain was obtained at a nearby hospital and showed the profound atrophy that, again, was highly consistent with, but not proof of, the disease. Even if she now had Alzheimer's disease, we wondered if she had the late-onset variety rather than the familial or early-onset form. Fortunately, we were able to obtain medical records on Hannah going back to her early fifties. These records confirmed that personality changes and difficulties with memory had indeed occurred in this decade and had been progressive, but crept at an unusually slow pace compared with that of others in her family. Still, there seemed no other explanation, and we accepted her diagnosis as a case of familial Alzheimer's disease.

Harry, a man in his late fifties, another great-grandson of Hannah but lost to the rest of his family for many years, was discovered by Charles to be living in a small Western town. His wife described to us what sounded like familial Alzheimer's disease and agreed to fly with him across the country. At Boston's Logan Airport, I met a frail gentleman in his late fifties accompanied by his wife of the same age helping to get him down the steps. It must have been a heroic effort for them to have made the trip. Had I known the extent of his impairment, I would have made other arrangements for his study. He was clearly in the late-moderate stages of Alzheimer's disease and was already developing the impaired movement and muscular rigidity that so frequently accompany this type of dementia. By the light of day, he remained sedate, resolute, and uncomplaining. Yet by night, in his sleep, and every night for the past two years, so his wife told me, she would be awakened by his primal and unanswerable scream "Why me, why me?" and his sobbing.

We were able to help Harry a little. We successfully treated his depression and improved his mobility enough so that his wife's wish that he could make one last vacation trip to his favorite part of the country became possible. Still, as with others, we could do nothing to arrest the progression of his disease.

Lucy, a great-granddaughter of Hannah, was a spry, vigorous teacher who prided herself, as I later learned, on her attractiveness and conviviality. Not long into the phone call that I placed to her in a follow up to my letter she told me, "Yes, I have it. I have had the earliest symptoms for two years." Would she come to our medical center for a full evaluation? "No, I fear flying even more than I fear Alzheimer's disease." Then a pause, "Yes, I'll come. Of course I'll come. I must—for the sake of my children."

Lucy was then 51 and still teaching, but with increasing

difficulty. Although she herself was convinced that she had the earliest signs of the disease, she insisted that we tell her nothing. Her symptoms and signs, though still mild, seemed definitive enough for a diagnosis, and so we told Peter.

We, the neurologists and neuropsychologists, were always kept in the dark ahead of time as to the DNA markers of each patient and as long as we harbored any uncertainty regarding diagnosis; in other words, we remained "blinded." Only after I gave Peter a definitive diagnosis—sometimes after several years of following a patient—would he tell me how the result affected the lod score. He informed me, "Lucy represents a recombination with respect to one of the markers." She "linked" with two markers toward the "centromeric" or middle end of chromosome 21 as Peter thought she might from previous studies, but she did not "link" with a marker previously thought to "link" with the disease toward the far or "telomeric" end of the chromosome.

Could we have made a mistake in Lucy's diagnosis and wrongly misdirected the attention of the molecular geneticists away from a correct location? After all, Lucy's symptoms were quite mild. She was still teaching. Several months later Peter told me that there were now two cases in other families that had confirmed the results we had found in Lucy's case.

A year and a half later, Lucy phoned. She had given up teaching and told me, "I know I could go on another year but it would not be fair to my students." She then asked me to confirm her previous self-diagnosis and requested that I submit our results so that she could receive disability and retirement benefits. Lucy asked, "Doctor, did I help the study?" I replied, "Yes, Lucy, you scored; you scored big." I tried to explain that her DNA sample set a limit on how far the abnormal gene must be from the centromere. Assuming that the abnormal gene in her family lay on chromosome 21, then the entire chromosome from a given marker to the end of the telomere could now be excluded from analysis.

We were still left with over 30 people to screen who reported to us that they were asymptomatic. We plotted their geographic locations on a map of the United States, and Brian O'Donnell and Joan Swearer tried to figure out the most efficient flight plans so that they could complete the screening as expeditiously as possible. Their excitement with respect to the research was mixed with apprehension and concern for those they would visit. They prepared to reassure their subjects, for these were people who had not come to us; we had asked if we could visit them. While O'Donnell and Swearer were ready to reassure whenever and however possible, at times, it was our subjects who reassured our psychologists that they were right to have come to carry out their tests. In San Francisco, one of Brian's subjects, a married father of two children, invited him out to dinner and a tour of the city. In Southern California, Joan set out to test an active woman in her early forties who assured Joan, "This may be the most important thing I do in my entire life."

By the middle of 1988 we had brought four new affected cases to the attention of the molecular geneticists, bringing to eight the total of definitely affected individuals in Hannah's family and increasing the number of individuals who could now be classified with very high probability as "escapees."

DNA analysis can be carried out on any "nucleated" cells—that is, those cells with a nucleus that contains DNA. We tried to locate any useful tissue from affected members of the family who had died in past years. Frozen brain tissue, if such could be found, would also have preserved "messenger RNA." The messenger RNA carries the "message" from the DNA to the cytoplasm so that the instructions of the nuclear DNA to make a particular protein can be carried out. Thus, in addition to studying the brain's DNA, we might later have the opportunity to see if the message from a suspected abnormal gene was either over—or under—expressed.

Charles had previously given us a list of where every autopsy had been done on members of his family; Elizabeth now undertook to provide the same information on members of her family. The foresight of scientists who had worked with these families in the past had prompted them to save frozen brain tissue on at least eight cases between the two families. Robert Terry had kept a number of these samples in his laboratory at the Albert Einstein College of Medicine before he moved to San Diego. He told us whom to contact at Einstein. When we tracked the samples down, we sadly learned that a freezer failure due to an electrical problem in the 1970s had caused the loss of the precious specimens. This search had ended.

In 1981 brain tissue from another affected member of Charles's family had been entrusted to a "brain bank" in Cleveland, Ohio. Inquiries followed. "Yes, I found it," Dr. Pierluigi Gambetti told me, and "I'll be glad to send a sample to Dr. Hyslop."

I can remember the inscription over the entrance to a pathology lab at Harvard Medical School, "Here the dead give up their secrets to aid the living." Now, eight years after the donor's death, a few grams of brain tissue would be thawed in Peter's lab, incubated at body temperature, and the instructions of the long-frozen DNA could once again direct the synthesis of messenger RNA and perhaps provide yet another clue! The newfound specimen, the remains of one of the afflicted descendants of Hannah, brought the total to nine for our DNA linkage studies.

But beyond the DNA and the messenger RNA, each case represented an individual, a family, often a parent with children who might be at future risk. From these people that David, Brian, Joan, and I screened, we gave Peter a list of the "affected," the "escapees," and the "indeterminates." That was what he needed for his calculation of the lod scores. But such narrow terms could not do justice to the response of our subjects and to their predicament. Carol, the sister of my patient Jeff, once wrote to me:

> This disease, for all its horrors and ironies, has brought our family many gifts as well (easy to say when I am not one of the affected): throughout my childhood and into young adulthood I was taught to live life with a sense of responsibility but with humor and compassion. Having had to face the possibility of joining my mother and my uncles and aunt on that horrific mountain top called "the family disease," I have learned the value of being able to function and contribute. My work as a psychotherapist has a direct relationship to the fears I have had to face in my life, but enough of this.
>
> I long for the day that my niece and nephews, especially, will be released from the ghosts of their ancestors.

For me, the experience of working with Hannah's descendants had been a moving one. I had grown too. As a medical student, I had long struggled to decide whether to pursue a research career entirely in basic science, even if my work bore no immediate relevance to clinical problems, or to bring the fruits of present-day knowledge to the benefits of patients in the clinic. I had seen my choice then as one between "to know" or "to cure." After some 15 years of doing only laboratory research, I had returned to the clinic. In addition to continuing my basic research on the visual cortex, I now spent about one-quarter of my time in direct patient care. Still, my clinical work had borne no relationship to my basic research interests. I realized now that there had been a faulty logic in how I envisioned my choice so many years ago. The best option was "to know so as to cure." In this pursuit a physician-scientist can achieve a unity of mind and heart that I had never known before.

Thus, with renewed vigor before this intensive screening phase of our work came to a close, I also tried to determine if we could find a common affected ancestor who might have been the progenitor of the disease in both Charles's and Elizabeth's family. If there had been a common affected ancestor, then Peter would be dealing with precisely the same mutation of an identical gene. That knowledge would permit us to combine the results from the two families without having to make additional assumptions. Moreover, if we could go back even one or two generations, we might find leads on other relatives who might have the disease and thus increase our chances of finding the abnormal gene sooner. Yet, we could go back no further in Hannah's family. The names of her parents simply were not known to her living descendants. If a grave could be located, perhaps the gravestone would list the names of her parents. But we knew that no gravestone had ever graced the site of Hannah's remains.

We were able to go back one generation further in Elizabeth's family. Her great-grandmother who had the disease arrived by boat in Philadelphia in 1904. The ship's documents were found which listed the names of her parents, but my search could go no further back.

Then I learned of the genealogical libraries and services provided by the Church of Latter Day Saints—the Mormons—the same group that had long been helping Mark Skolnick and Gary White in their efforts to generate linkage maps. Their main resources were in Salt Lake City, Utah. I reached Tom Daniels, the head of the library division at the Church, who arranged for the head of their section on Eastern Europe and the Soviet Union to search their records for us. They searched their own records and the old "consular records" from Czarist times that had once been confiscated from Russian embassies and kept in trust in Canada. After an exhaustive effort, Daniels phoned and regrettably reported that no further trace of either family could be found.

And so my search for other relatives of Charles and of Elizabeth was over. I had looked everywhere possible. Everywhere except for Russia.

Was Russia possible?

Russia Once Again

In November 1987 I asked Professor Joseph Tonkonogy, the distinguished behavioral neurologist who had left Leningrad's Bechterev Institute in 1979 to come to the United States, whether there was any way of contacting neurologists in the Bobruisk area to learn if there might be an unusually high incidence of early-onset familial Alzheimer's disease in this region. Perhaps there would even be descendants of common ancestors of the two Byelorussian families that we were now studying. "Possible, but for Alzheimer's disease you want to contact Leningrad and Moscow." He then explained that Alzheimer's disease was present in all ethnic groups in the Soviet Union and was a matter of great interest there. Russia's research into Alzheimer's, unlike that in many areas of medicine, was on a par with that of the Western world. He volunteered to check in the medical school library for recent Russian publications.

A day later he returned with a paper on Alzheimer's that had just come out by Professor Svetlana I. Gavrilova, a psychiatrist in Moscow. In the Soviet Union, Joseph explained, Alzheimer's was diagnosed and managed by psychiatrists, not by neurologists as in the United States. Well then, I would write to Gavrilova. "No," said Tonkonogy, "you can't do that." All requests had to be forwarded to the Institute directors, he said. No individual physician or scientist in the Soviet Union could decide to enter into a collaborative study in general, let alone with someone from another country. And so off went a number of letters. Professor Modest Kabanov, director of the Bechterev Psychoneurological Institute in Leningrad, regretted that there were no useful pedigrees in Leningrad at this time, although he assured me that a collaboration with scientists in the United States would have been most welcome. Two Institute directors from Moscow sent similar regrets. One also noted that it would be unlikely that we could find relatives of the American families that we were studying still alive in Byelorussia "because of the genocide practiced there during World War II."

The response from Professor Marat E. Vartanyan, the acting director of the All Union Mental Health Research Center in Moscow, was positive. There were 21 pedigrees under study by Institute scientists, in which there were at least two living affected members.

I checked again with Peter Hyslop as to whether the search for new pedigrees should go forward. Again, he resounded, "Absolutely." In fact, since the publication of the breakthrough paper in February 1987, our center had identified no new large pedigrees—with the important exception of Elizabeth's family. Certainly, we and many other centers frequently encountered familial Alzheimer's disease. A patient or a spouse often reported that one of the patient's parents or grandparents had also had the disease, but that they were no longer living. Rarely did our patients know the medical histories of their first cousins and virtually never of second and third cousins. Even in the few small new pedigrees that we were establishing, there had rarely been an autopsy to establish Alzheimer's disease as the definitive cause for the dementia within the family. In fact, virtually all the pedigrees subject to DNA linkage studies in the 1986–1987 period had been well described in the medical literature prior to 1980. It could easily take five to ten years to document a useful pedigree thoroughly. It was time that mattered. The Western world's literature had already been scoured. Families in Scandinavia, subject to classical studies in the 1950s and 1960s, were still under medical scrutiny but had few living affected members. Families from England and Western Europe were already under study, but many more would be needed. Thus, Peter encouraged me to follow up on the pedigrees in the Soviet Union.

Thrilled as I was to learn that the Moscow workers might have useful pedigrees and had pledged full cooperation, I soon became disheartened when I read an article in the February 5, 1988 issue of *Science*. The story focused on the abuses of dissidents's rights by psychiatrists in the Soviet Union and showed a photograph of Vartanyan and Andrei V. Snezhnevsky standing together some years ago.[1] Snezhnevsky had been the most recent leader of the Moscow School of Psychiatry. During the first decades of this century, his predecessors hypothesized a form of schizophrenia called sluggish schizophrenia. With time, however, the term became a notorious pretense, a perversion and mockery of medical values, under which the KGB abused the rights of political dissidents and human rights activists. The abuses during the 1970s and early 1980s were especially blatant. Snezhnevsky himself had been a superb clinical psychiatrist, dedicated to patient care and building a Psychiatric Institute of the highest quality. He was said to be in the clinic by 8 A.M. seeing patients. He sought the most capable people for each position. He alone among all directors of top psychiatric institutions in Russia defied the official government policy that prohibited the hiring of Russian Jews to top positions. But he did not defy the KGB when it came to the issue of the dissidents.

Snezhnevsky was dead now for almost two years, and Vartanyan was his acting successor. The *Science* article condemned Vartanyan as well and urged his removal from power before Soviet psychiatry could again enter the World Psychiatric Association. Such was the position of the American Psychiatric Association. I quickly showed the article to Tonkonogy. "Vartanyan is a biochemist interested in schizophrenia," he told me. "He doesn't see patients. He has never committed a dissident." Another prominent physician, Dr. Bernard Lown, a founder of the (International Physicians to Prevent Nuclear War) who knew Vartanyan through this organization, told me that Vartanyan's own father had been killed by Stalin and his mother condemned to a gulag for most of his childhood, leaving the young Vartanyan and a brother to fend for themselves for years in an environment that was then decidedly hostile to them.

Tonkonogy told me the article had been correct about Georgy Morozov, however. Morozov was the chief forensic psychiatrist in the Soviet Union. He indeed bore direct administrative responsibility for the abuses of dissidents in the Soviet Union by psychiatrists which had been revealed to the Western world by many heroic Soviet psychiatrists. Among them was Dr. Anatoly Koryagin, who himself had been imprisoned and tortured for opposing the heinous practice. But was Vartanyan accountable for the abuses? I contacted Ellen Mercer, the spokesperson for the American Psychiatric Association, and she concurred with Tonkonogy that Vartanyan had not himself abused political dissidents. However, she stated that at international meetings he had denied that such abuse had occurred. Should I drop the collaborative opportunity on Alzheimer's disease just getting started? No, she advised. Although the APA could not and would not recognize Vartanyan officially, the APA encouraged contacts between individual American scientists and individual Soviet scientists, especially on worthwhile projects.

On March 1, 1988 the Soviet Union officially outlawed and made it a crime for anyone to be committed to a psychiatric institution for political beliefs or for any reason other than a valid need for psychiatric care. The "special" psychiatric hospitals which the KGB controlled in the past were now transferred to the mental health authorities.

In April, Vartanyan invited me to visit his Institute in June, but a 10-day delay in the receipt of the official invitation by telegram precluded my making plane reservations in time. Our meeting was reset for early July, at the close of the Reagan-Gorbachev summit on nuclear arms reduction.

Two weeks before I left for Moscow, Jeff returned to my clinic for a followup visit. I had been carefully adjusting his medications so that he would retain some useful mobility as the muscle rigidity of the disease unrelentingly advanced. Jeff could attempt a little work in his yard now, but that was about all. I tried to explain to him why I'd soon be going to Russia, that I would also ask the Russian doctors to search for members of his family there, but I doubt that Jeff could still understand. There was essentially no hope now of helping Jeff further. I could only hope that the results of my trip might in some way speed the day when we could help others.

The Twenty-one Pedigrees of Nina Ivanovna Voskresenskaya

My wife Linda and I were met in Moscow on July 17, 1988 by Vera Germanovna Kukushkina, a charming, vibrant, intelligent young woman of 24 who was to be our interpreter. Her father was a prominent scientist in a medical genetics institute in Moscow, and her mother was a teacher of music. As would soon be apparent, Vera's cordiality, openness, and eager excitement set the tone for the four days that were to follow.

After a quick check-in and change of clothes at the Rossiya Hotel, we were off for a working lunch at the All Union Mental Health Research Center. We were greeted by Professor Vartanyan, now the Institute director, and Pyotr Morozov, an Institute psychiatrist and Vartanyan's assistant. I quickly realized that Vartanyan was an extraordinarily intelligent and urbane man in his mid-fifties, deeply committed to solid research in biological psychiatry. Both spoke English flawlessly. Pyotr Morozov, not to be confused with Georgy Morozov, graciously co-hosted the opening session.

The All Union Mental Health Research Center (more or less

the Russian equivalent of the intramural branch of our National Institute of Mental Health in Bethesda) was a psychiatric research institute with patient care facilities as well. Highlighting the more formal afternoon session were two talks by members of the Research Center, one by Professor Svetlana Ivanovna Gavrilova on the prevalence of Alzheimer's disease in the Russian republic and the other by Dr. Nina Ivanovna Voskresenskaya on familial Alzheimer's disease in Russia.

Dr. Gavrilova began to speak in what must have been an arduously prepared but polished English text. Her findings on the incidence and prevalence of Alzheimer's disease as a function of age in the Moscow region were essentially identical to those recently published for populations in the United States. Her estimates of the increased risks of first-order relatives—that is, of a sibling or a child—of those affected with Alzheimer's, compared to an age-matched sample of the general population in Russia, were virtually identical to results from the most recent studies coming out of North America. Voskresenskaya spoke next. It was she who had over the past five years assembled the 21 pedigrees of early-onset Alzheimer's disease.

Later that week Svetlana, Nina, and I scrutinized the 21 pedigree diagrams for information needed in our work. The hand-drawn charts representing the family trees used the worldwide accepted conventions for representing male and female, affected and unaffected, living and diseased. Thus, I could quickly assess the genetic informativeness of each family record. These pedigrees had been developed not for the purpose of future molecular genetic studies but rather for determining the risk of first-order relatives acquiring Alzheimer's disease. Many of the families had only one or two living members with the disease and not enough at-risk individuals who had passed the age of risk for the families to be of likely use now. Seven pedigrees appeared particularly useful because they contained three or four living, affected members and a number of escaped family members who could serve as controls. Several other pedigrees, though not as informative, would some day be useful in addressing the issue of possible genetic heterogeneity—that is, the question as to whether the defect of more than one gene could cause familial Alzheimer's disease of the early-onset form. As soon as a first defective gene for any genetic disease is discovered, it becomes a relatively simple matter to see if the same mutation is present in other families—even if there are only one or two living affected individuals. A number of the pedigrees were verified by autopsies, proving that the dementia in question was indeed

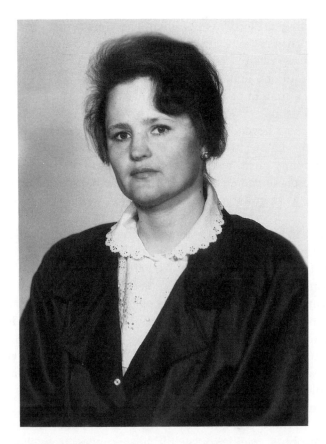

Dr. Nina Ivanovna Voskreskenskaya. (Courtesy of Dr. Voskreskenskaya.)

Alzheimer's disease and thereby greatly enhancing the value of the pedigree.

The Russian pedigrees were not likely related to the two families of Russian-Jewish origin that we were studying in the United States since only one of Nina's pedigrees had Jewish family members. Nina volunteered to see if the ninth branch of Hannah's family could be found. If this branch also carried the abnormal gene, the discovery of even a few new cases could be profoundly important for us. But the search, we were told, would be a difficult one.

The night before we left Moscow, after an evening of ballet, Nina said good-bye at our hotel entrance, tenderly holding out a manila packet that told of the lives and fate of her 21 pedigrees. "These are like my children," she said softly, as she entrusted the packet into my hands.

We were an odd group that boarded the Air France flight to

Paris: smartly dressed businessmen who were returning to Western Europe with joint ventures planned, Russian citizens taking their first but long-awaited trips to Western countries to visit relatives, summer tourists, and my wife and I guarding a box of filled test tubes and a manila packet that would not leave our hands until we reached Boston.

The plane headed northwest, and in time followed a long river, perhaps the Dvina, which led us to the Baltic Sea. The afternoon was incredibly clear, and from 30,000 feet the view was extraordinary. The plane veered westward over the Baltic. In less than four hours, we were in Paris, and from Paris we went on to Boston.

How had it happened that Russia, which technologically was far behind the West in many areas of medicine, had developed such a strong tradition in the field of Alzheimer's research that its scientists could provide help in this project? And how had it happened that Nina Ivanovna, as a young medical doctor and scientist, had developed such a driving interest in Alzheimer's disease? In a sense the answers to these questions go back to Kraepelin and Alzheimer. Erich Sternberg, a medical graduate of Humbolt University in Germany, was schooled in the classical traditions of German neuropathology with twin interests in Alzheimer's disease and schizophrenia, but in the 1930s, as a German Jew, he had to flee his country. In these years, however, the gates to America were almost closed; only a trickle of would-be escapees from Hitler were allowed to enter the United States. The stories of the prominent emigrés who fled to the safety of America and England, when they could, are well known, but Sternberg fled to Russia in 1933. He immediately became involved in clinical and research activities and worked with famous Soviet psychiatrists such as Andrei Snezhnevsky.[1] Tragically, Sternberg and his wife were arrested in 1935, accused of being spies for Germany, and were sentenced to a labor camp. Yet even in confinement, Sternberg continued a clinical practice.

In 1953, with the death of Stalin, Sternberg was released. Snezhnevsky sought out and found Sternberg and invited him back to Moscow in 1956. Eventually, Sternberg, even after having been deprived of 20 of his best years for research and teaching, returned to his position at the Institute of Psychiatry at the Academy of Medical Sciences of the USSR (then the All Union Mental Health Research Center) in Moscow and lived out a long scientific career making many contributions to the fields of Alzheimer's disease and schizophrenia. He trained Svetlana Ivanovna Gavrilova, who now was in the midst of her own distinguished career. She, in turn, trained Nina Ivanovna Voskresenskaya, who had begun her work five years ago. So it was that the first blood samples of the Russian pedigrees had come to Boston, in a sense handed from Kraepelin to Alzheimer, to Sternberg, to Gavrilova, to Voskresenskaya, and now to Hyslop.

How Many Genes?

By midsummer 1988, the tacit assumption that all cases of familial Alzheimer's disease might be caused by various mutations within a single gene was to be challenged head-on. No one yet knew what fraction of Alzheimer's cases was caused by a single dominant genetic defect, or how many different genes might be abnormal in various families where an inherited defect was virtually certain, based on the pattern of inheritance of the disease over several generations. Probably only 10 to 15 percent of all cases of Alzheimer's disease occur with a pattern of inheritance suggesting a single dominant genetic defect as clearly as that in Hannah's family.

Estimates of what fraction of all Alzheimer's cases are genetic in origin vary enormously. Part of the reason is that the incidence of the disease increases greatly in the older age groups, rising some 400-fold from age 40 to age 80. By conservative estimates, the incidence of the disease by age 85 is at least 15 to 20 percent, with several studies suggesting an incidence at that age as high as 50 percent. If the majority or even some of these late-onset cases are sporadic in origin and do not reflect a genetic basis, then it becomes very hard for neurogeneticists to study late-onset Alzheimer's disease even within a single family and to determine the mode of inheritance with any certainty.

Even so, a number of large-scale studies suggest that an individual's risk of developing Alzheimer's disease later in life, after 70 years of age, is perhaps three times greater if that person had a parent or other first-order relative, such as a sibling, who developed late-onset Alzheimer's disease than if such relatives lived to late ages and did not develop the disease. Apparently, these increased risks are real; it is less clear whether the risks represent truly genetic factors, common susceptibility to environmental factors, or lifetime dietary habits. Thus, the etiology of late-onset Alzheimer's disease, even if partially genetic in origin, may have to await either the resolution of the genetic basis of the more clearly definable early-onset form or the location of a gene in a few large families with a clearly defined late-onset pattern of inheritance.

Efforts to identify large numbers of families with familial Alzheimer's disease occurring in midlife began in earnest in the United States in the mid-1960s with the work of Leonard Heston and his colleagues. During the 1970s Robert Cook and his colleagues published a review of many of the then known well-documented kindreds with familial Alzheimer's disease. Workers like Heston and Cook were always on the lookout for factors that were common for the various families as well as those that might distinguish the expression of the disease in the different kindreds. During the 1980s Thomas Bird, Gerard Schellenberg, and their colleagues at the University of Washington's Alzheimer's Disease Center in Seattle identified a group of more than 80 families with well-documented histories of dementia and selected a smaller subgroup of 24 kindreds for molecular genetic research.

In the course of documenting the origin of their pedigrees, the Seattle group realized that three of the families described by Cook and two additional families that they had studied were descendants of a small group of settlers whose ancestors had emigrated from two adjacent villages in the Southern Volga region of Russia. Members of this group comprised the American descendants of a remarkable cultural group known as the Volga Germans. In 1762 Catherine Empress of Russia, herself of German birth, invited settlers from Western Europe to settle and farm the open plains south along the river Volga to curtail the series of invasions by marauding tribes from Central Asia.

F. C. Koch[1] described the history of the Volga Germans, and the Seattle group summarized the relevant history with respect to that small fraction of their population that became afflicted with familial Alzheimer's disease. Probably some 90 percent of the settlers who accepted Catherine's offer originated from Germany; others came from Denmark, Holland, Belgium, and France; and still smaller numbers came from other European countries. Most of the German settlers emigrated from the Hesse-Frankfurt area, which had been especially depressed because of the frequent wars and famines during the preceding century. Eventually, during the 1760s some 27,000 Germans migrated to Russia, including about 500 who settled in the town of Frank and about 400 who settled in the town of Walter, some six miles away. The Volga German group retained their German identity, language, and customs and remained geographically and socially apart from the rest of the Russian population.

As Bird and his colleagues note, however, the social and political turbulence of the World War I era led to hard times for the Volga Germans. Not only did they lose their land and have to join the Russian Army, but they were also accused of being German sympathizers. As a result, many of Volga Germans headed for the United States; they settled mostly in the Midwestern states of the Dakotas, Kansas, Nebraska, and Colorado, as well as the West Coast.[2] The Seattle group estimated that by the mid-1980s there were some 300,000 American descendants of the Volga Germans.

Bird's group suspected that the five families of Volga German ancestry afflicted by familial Alzheimer's disease might all be descended from a common ancestor, an example of what is called the "founder effect" in genetics. Yet the Seattle group was aware that the assumption remained unproved, and their own research on the five families established an almost 15-year spread in the age of onset of Alzheimer's disease in the various families. In some kindreds, the disease often began in the mid- or late forties, with the mean age of onset at about 50, which was not very different from that in the descendants of Hannah. Other Volga German kindreds, however, revealed a mean age of onset as high as 63 years.

The Seattle group noted that there was no known relationship between the families they studied and the two Byelorussian Jewish families represented by the families of Elizabeth and of Charles. As they concluded, "intermarriage between these latter kindreds and the Volga Germans is unlikely because of a strong adherence to separate ethnic identities and a geographic separation of over 800 kilometers." Assuredly, since the Volga Germans migrated in the 1760s, the possibility that the two groups shared a recent common ancestor was virtually impossible, but the possibility that the two groups shared a remote common affected ancestor with an Alzheimer's disease gene could not be excluded.

Yet if the Seattle group could be certain that their five families were descended from a common ancestor, their search for an abnormal gene would be much simplified. Genetic analysis of the kindreds would have been vastly facilitated if the precise geneologic relationships between the various kindreds could have been ascertained. Even in the absence of knowledge as to whether the various Volga German pedigrees were related, however, the Seattle group had enough definitive information within each of the separate families to test for genetic linkage in up to seven different Volga German kindreds and in another eight families of non-Volga German ancestry. Thus, when Peter Hyslop's paper reporting linkage to chromosome 21 was published in early 1987, the Seattle group was ready to begin testing to see whether the two markers linking familial Alzheimer's disease to chromosome 21 in our group of families also linked to their families in the Western states.

The results of the Seattle group were unequivocal with respect to one of the previously linked markers, the one called D21S1. In both the Volga German families and in the Seattle group's other families, the marker failed to link to Alzheimer's disease. In their studies of Volga German families, the Seattle group could rule out an abnormal gene within over 15 centimorgans (about 15 million base-pairs) from the D21S1 marker. The results from their other families were against, but only weakly against, linkage to the same D21S1 probe. The Seattle group also failed to find any positive evidence for linkage of the disease to the D21S16 marker, which our group believed was linked to familial Alzheimer's disease at least in the Italian family from Calabria. With respect to the D21S16 probe, the Seattle group's evidence against linkage was far less conclusive.

The Seattle workers pointed out the possible interpretations of the discrepant results. First, they realized that their findings provided strong evidence for genetic heterogeneity. That is, more than one gene likely caused familial Alzheimer's disease, assuming, of course, that the initial report of our own group was correct in linking the disease to a locus on chromosome 21. The Seattle workers also realized that the average age of onset of the disease in their families was significantly higher than in some of the families reported by our group, and they raised the possibility that a gene on chromosome 21 might account only for very early-onset cases of familial Alzheimer's disease. Finally, the Seattle workers noted another possibility that could not be ruled out—"the possibility that the positive lod scores obtained by St. George-Hyslop et al. were a chance occurrence." But the Seattle workers, in agreement with several previous reports, also excluded defects in the amyloid gene as the cause of Alzheimer's disease in several of their families.

Meanwhile, Allen Roses and his colleagues at Duke Medical Center had been studying a number of rather small families with late-onset (onset of greater than 60 years) Alzheimer's disease presumed to be familial in origin. The Duke workers initially failed to find linkage of Alzheimer's disease to the D21S1 region on chromosome 21, although in later reports they did report evidence in favor of such linkage in some families with early-onset Alzheimer's disease.

By now, in late 1988, we could not be certain how many different genes might be causing familial Alzheimer's disease in the various families. The same problems that had plagued the search for Alzheimer's disease genes several years earlier still hindered all research groups. In almost all cases, there simply were not enough definitely affected individuals within any given kindred either to identify or rule out a particular gene locus, given the relatively widely spaced DNA probes then available. In the absence of obvious candidate genes, such as the amyloid precursor protein gene that could be readily tested, future progress still depended on the advancing technologies for discovering DNA probes for ever more closely spaced regions of the human genome and on the search for more definitely affected members within the various pedigrees.

From my point of view, the situation was especially disturbing. The results of our group in establishing linkage of familial Alzheimer's disease to chromosome 21 had been dominated by the strongly positive results from the Italian family originating in Calabria. But this pedigree was in many respects the most atypical of all those considered to fall under the rubric of familial Alzheimer's disease. For example, disease onset was especially early, with a mean age in the mid-thirties. Moreover, seizures—which are not typical of Alzheimer's though they may occur in late stages in a small percentage of patients—developed in the majority of patients of Calabrian ancestry, often in very early stages of their disease. And there were a variety of pathological findings in the brains of these patients that were not typical of the early descriptions of Alzheimer's disease.

Yet in most respects, the clinical and pathological findings in the Italian family qualified it as an example of familial Alzheimer's disease. Moreover, the statistical evidence was strong, though not definitive, that a genetic defect was located on chromosome 21. In addition, the lod scores for Hannah's family and the British Canadian family for linkage to the D21S11 marker remained positive, albeit weakly so. Thus, the possibility remained open that the genetic defect in these two families involved the same gene but not the same mutation as in the Italian family.

Taken on their own, the results in Hannah's family and in the British Canadian family could not make a case for linkage of their genetic defects to chromosome 21. Could the genetic defect(s) in these two families reside on another chromosome? There was still no way to settle this issue. Recall that for many genetic diseases all known defects have linked to a common site such as the short arm of chromosome 4 in Huntington's. Similarly, all cases of Duchenne's muscular dystrophy are caused by abnormalities within the "dystrophin" gene on the X chromosome, although a large number of different disease-causing mutations of this gene have been found.

The persistent dilemma we faced was not the fault of geneticists. Each had followed the time-honored rules for carrying out linkage studies by the family method, and when there was no single affected family available to settle an issue, then several families, presumably with the same disease, had to be studied together. But what if even the familial form of Alzheimer's disease was not a single disease entity but a handful of diseases of different primary cause which produced a basically similar clinical and pathological picture—perhaps even by eventually affecting a final common biochemical pathway? Suppose instead of Alzheimer's disease, we should have been thinking of Alzheimer's *diseases*?

Ideally, parallel tests in both Elizabeth's and Hannah's pedigrees and in the British Canadian family for linkage of Alzheimer's disease to sites on other chromosomes would have been desirable. Again, however, the number of affected individuals within any given family was not then sufficiently large to convince molecular geneticists that they might find a definite answer were they to begin the arduous testing of other chromosomes. If we were to make more rapid progress on the genetic defect in Hannah's family, we would have to discover a way to increase the informativeness of the pedigree.

The Third Dove Goes Forth

The early autumn of 1989 combined conflicting emotions of hope and dread—hope attending the announcement of a potentially major discovery and the dread oppressing a physician when he must helplessly witness the rapid decline of his patient. In the September 21, 1989 issue of *Nature*, Dennis Selkoe and his co-workers, Catherine Joachim and Hiroshi Mori, reported that the same amyloid β-protein found within the senile plaques and blood vessels of brains of patients with Alzheimer's disease could usually be found within skin samples from patients with the late-onset form of the disease.[1] Selkoe had less frequently found these deposits within the skin of elderly normals; he did not yet know whether the deposition of amyloid in the skin would distinguish Alzheimer's victims from normals in middle-aged individuals.

In recent years, several groups had claimed that Alzheimer's disease was not exclusively a disease of the brain. Of all such studies, the report of Selkoe's group was especially tantalizing because amyloid, the substance perhaps most closely associated with the pathology of the disease, had now been found in tissues outside the brain. Did a common precursor substance reach the multiple organs through the bloodstream as in some recently understood "systemic amyloidoses"? Or did the deposition of amyloid in multiple tissues suggest that the protein may simply be produced locally within each?

The discovery of Selkoe's group once again reminded us that the amyloid gene, the gene directing the synthesis of the precursor of the amyloid protein, was located only about 10 million base-pairs away from the suspected site of a defective gene in familial Alzheimer's disease. We strongly believed that the amyloid gene itself was normal in Alzheimer's disease—at least in the patients with whom we were working. However, was the relative proximity between the two genes sheer coincidence, or might the fundamental abnormality of familial Alzheimer's disease be manifested through an alteration of the activity, or "expression," of the amyloid gene?

We still did not know any more than Kraepelin and Alzheimer had, whether the appearance of amyloid was a causal factor in Alzehimer's or simply a consequence of a more fundamental abnormality. Nevertheless, the possibility of studying part of the disease process in an organ as accessible as the skin brought much promise. The next obvious question was whether the amyloid deposits would be found in the skin of patients with early-onset familial Alzheimer's disease who were much younger than the patients that Selkoe's group had first studied. They had not yet found amyloid within the skin of any normal subjects under age 70; thus, the finding of amyloid in the skin of patients with early-onset Alzheimer's disease who were still in their mid-forties to mid-fifties might be of special significance. At the very least, if amyloid were found in the younger subjects, it perhaps occurred independently of the normal process of aging.

Thus, Dennis Selkoe invited our group to join him in a study of members of Charles's and Elizabeth's families. By early October, I submitted an application to our Committee on the Protection of Human Subjects asking permission for us to obtain small skin biopsy samples from our affected patients, from subjects who were neurologically well but still at risk, and from family members who had escaped the disease. Even though the procedure carried no known serious risk, any such study requires careful medical and ethical considerations, and we knew that we would have to wait four to six weeks for the committee to review and pass on our proposal.

One of the patients we would want to test was Jeff. When I had first met him some four and one-half years before, he was still well enough to assert his conviction that he wanted to participate in research to help others in every way possible. As his illness progressed, and as long as he was capable of expressing himself, Jeff pleaded that he wanted his family and physicians to respect his wishes in two matters. First and above all, he did not wish to be a further burden to his family in the period that he knew was approaching. Second, he realized, too, that his disease would soon rob his life of any meaningful existence. When that time came, he did not want any medical efforts made that would simply prolong the process that had robbed him of mind and tore without respite at the hearts of his family.

By the early fall of 1989, the time that Jeff had long dreaded had now come. For the past several months, he had not been able to speak beyond single words or to feed himself. Several physicians carried out careful neurological evaluations and could find no cause for his rapid decline other than Alzheimer's disease. A touch of pneumonia gave hope that a mercifully quick end to his suffering might come soon, and Jeff was transferred to a hospice for compassionate terminal care. Under Massachusetts laws, a physician is permitted to respect the wishes of a patient or the substituted judgment of the patient's family, and is not compelled to prolong the process of dying by treating pneumonia in a patient who is gravely ill for other reasons.

We had not yet received permission to obtain skin samples from living patients, nor would we have wanted to do any procedure on a patient as ill as Jeff. Jeff's family agreed, that upon his death, a skin sample could be taken at the time of autopsy. Selkoe, aware of the urgency of the situation, sent me by overnight mail his protocol for preparing skin samples. When I returned from the clinic late Monday afternoon on October 16, Selkoe's packet had arrived. I read the material carefully and with a slight sense of foreboding left the packet on top of my desk. When I reached home, I found a message from the physician at the hospice stating that Jeff had passed away at 5 P.M. even as I had been opening Selkoe's packet. Before Jeff's death, his wife had reminded me of Jeff's conviction, which she fully shared, that all possible opportunities for research should be grasped. By prior arrangement, everything had been done to ensure that precious samples of tissue could be processed or frozen as soon as possible for later studies. Jeff's funeral was held two days later, on a day somber and gray, fitting, it seemed, for the occasion.

And here was Charles, at yet another funeral, this time for his nephew. With only three hours left before boarding his plane for his return trip to the Midwest, it was he who would ask if there were enough time for him to provide a skin sample as a normal control for Selkoe's studies. Charles did not have to be asked.

Although there was a long way to go, the pace was quickening now. That very day, our neuroanatomist, Jim Hamos was sending small samples of skin to Selkoe, whose group would soon discover deposits of amyloid within Jeff's skin.[2] For Peter Hyslop and our neurogenetic group, a bit of brain tissue frozen soon after death would later offer a priceless opportunity to search for alterations in messenger RNA. This material might give further clues of the basic genetic defect and how the genetic defect "expresses" itself in the brain itself.

For three generations now Charles's family had waited for solace, just as Noah had sent forth doves in search of dry land. When tissue from Jeff's grandmother, Ida, had been sent forth some 50 years ago, the news returned that she had Alzheimer's disease, but prospects for any understanding of its causes, let alone treatment, were almost beyond human comprehension. The seed that Charles had helped sow some 30 years before, when he had encouraged Saul Korey, Robert Katzman, and Robert Terry to study a sample of tissue from the brain of Sarah, Jeff's mother and Charles's sister, had taken root, and with it a generation of mature investigators had made a science out of the study of Alzheimer's disease.

Now the third dove had been sent forth. Perhaps this time, new land might soon be in sight.

PART IV

The Path Ahead

Lost Tribes

In the early spring of 1989, Nina Ivanova Voskresenskaya journeyed from Moscow to Dnieperpetrovsk in the Ukraine in search of the Russian descendants of Hannah and Shlomo. We knew that Hannah's daughter, Rosa, who had remained in the Soviet Union had borne seven children of whom six—one son and five daughters—had remained in the Ukraine. Rosa's second son, David, immigrated to the United States in 1912; he lost contact with his family in the Soviet Union during the period of forced collectivization and famine in the mid-1930s. Even so, David held on to the single sheet of paper bearing the last known address of his sister Lisa.

In late middle life David developed the neurodegenerative disease amyotrophic lateral sclerosis, sometimes called Lou Gehrig's disease after the late great baseball player who succumbed to it. In this disease, the loss of nerve cells in the spinal cord leads to paralysis of motor function. The possibility of a relationship between the cause of this degenerative disease and Alzheimer's, which has been reported to occur together, albeit rarely, in some families, provided us with an added incentive to find David's family.

Before he died, David passed along the address to cousin Bryna, the family historian, who had held it since 1956. With this address in hand, Dr. Voskresenskaya found the apartment building of David's last known relative, but none of the present residents had any recollection of Lisa T., who had lived there over a half century ago.

Inasmuch as the five sisters—if they had survived—would probably have married long ago and been living under the surnames of their husbands, Nina Ivanova had little chance of tracing them directly. Instead, she made an extensive search of the records of hospitals in the district, alert for any cases of early-onset Alzheimer's disease; then she worked backward to see if any of the victims matched the names of the five sisters that we had given to her. Once again, not a trace was found.

Yet there was one male with the family name who might be traceable. After much effort, Dr. Voskresenskaya found the address of an apartment building where Beryl T. was reported to have lived in the 1930s. She visited the apartment complex but could find no trace of the family. Finally, as she was leaving, Nina Ivanovna came upon a woman in her late eighties who told her that she remembered the family, "Oh yes, Beryl, his wife and their four young children fled toward Central Asia in July 1941 as the Germans were approaching." But to her recollection, neither Beryl nor his family had ever returned.

Thus, Nina Ivanovna's search for the descendants of Hannah and Shlomo in the Soviet Union had come to end and, with it, the chances that in the near future we might find other cases of Hannah's descendants with familial Alzheimer's disease and thus hasten our linkage work.

When I first met Jeff, lymphocyte cultures for DNA analysis could be established on four living affected descendants of Hannah, including Jeff. In the four years since his visit, we had identified four more cases in other members of his family. After further search, we had obtained tissue from a brain bank of another affected member who died in 1981. This brought the total of DNA samples from Alzheimer's victims to nine.

Nine was certainly a very meaningful number for genetic linkage analysis, but far from the hundred or so cases of Huntington's disease that Nancy Wexler and her colleagues had found in Venezuela. For us, every single lead was important to follow.

Eagerly, I decided to ascertain the clinical status of fourth-generation descendants of Hannah whose at-risk parent had died without living past the upper limit of the age at risk. In the course of talking with one descendant, Dr. N., an epidemiologist living in Texas, I learned that he was fully well; but quite by accident, I fell upon another lead. Dr. N. had served in the U.S. Public Health Service at the National Institutes of Health in the mid-1950s. One day while making rounds at NIH's clinical center, he discovered that two of his aunts were patients there awaiting brain biopsy for diagnostic purposes. Both biopsies were diagnostic of Alzheimer's disease. With help from colleagues at NIH, we tracked down the old medical records of his aunts, but unfortunately no brain tissue had been saved.

Then I recalled the 1981 paper from NIH by Goudsmit, Gajdusek, and their colleagues in which they had injected brain tissue from affected members of both Charles's and Elizabeth's families into mammals to see if the disease was transmissible. We contacted Dr. Gajdusek to see if he had saved such tissue and would share it with us. He remembered the two families quite well but, given the enormous number of brain samples he had worked with over the years, he could not remember the names of these two donors. The computer retrieval system depended on having the first and last names of each patient.

Then I thought of running the names of every patient in Charles's and Elizabeth's family who had acquired Alzheimer's disease in the period between 1950 and 1988 through Gajdusek's file system. Dr. Marc Godek, a young scientist starting out in Gajdusek's lab, ran the search for us and excitedly called me to announce that he had found frozen brain tissue on two individuals: one in Charles's family and the other in Elizabeth's. Ample tissue was available in the first case, but only 2 grams remained in the second. (For a sense of this amount, consider that a nickel weighs approximately 5 grams.) Gajdusek, his colleague Joe Gibbs, and their lab groups met to decide whether to share any of the 2 grams of tissue left with us. They all agreed to do so and sent a vitally important 1 gram of cerebral tissue to Hyslop at the MGH labs. Now we had readily testable DNA samples from ten individuals in Charles's family and from four in Elizabeth's.

Knowing of the freezer failures in other labs where tissue from affected patients had once been deposited, we knew that there was no prospect of finding any other frozen tissue. For a time, it seemed that we could only wait for more cases to develop among the at-risk group we were following. But, in February 1988, Charles would again change the course of our thinking. Our medical center had arranged for Jim Gusella and Peter Hyslop to present lectures on their respective searches for the Huntington's and Alzheimer's disease genes, and Charles was on hand to hear the updates. When we found each other in the hall, he handed me a small, tightly sealed test tube containing what must have been less than a gram of a white crystalline substance. The substance was DNA that he had extracted from the formalin-fixed brain tissue of a deceased affected member of his own family.

Charles explained the then newly developing PCR (polymerase chain reaction) technology that was revolutionizing molecular biology by making it possible for scientists to amplify by the millions and billions a specific stretch of DNA from even a single cell. Suppose that a scientist has acquired a minute quantity of DNA from a victim of a genetic disease and must determine whether the defective gene links to either the A or B form of a test marker. But suppose also that the quantity of DNA available is so small that the required measurements fall below the sensitivity of existing techniques. It would be ideal if the geneticist had available some preliminary method that would "amplify" or make multiple copies of all the strands of DNA in his precious sample, but no such general method exists. However, thanks to the invention of a remarkable technology by K. B. Mullis,[1] a scientist can amplify a preselected sequence of DNA and, if enough of the original DNA is available, analyze other DNA sequences, if needed, later.

The PCR technology requires that the scientist know beforehand the unique sequence of a band of nucleotides—called the primers—at both the top and bottom of the much longer strip of DNA that has been targeted for amplification. The scientist then adds an enzyme called DNA polymerase to the mix of the original DNA and the primers, and the enzyme directs the rapid synthesis of copies of the preselected strip of DNA. With each cycle of synthesis, the number of chosen DNA strands will double. When enough DNA has been synthesized, the geneticist can use a bacterial enzyme to cleave the DNA at selected sites and determine whether the A or B fragment co-inherits with the disease. By successive reapplications of such methods, the geneticist edges ever closer to the defective gene.

Later, I handed the precious tissue to Peter. We both stood in awe of Charles, a man so committed to finding the cause of Alzheimer's disease that he would extract the DNA from a specimen of saved brain tissue from his own long-departed sister.

Although Peter had been well aware of the advances in PCR technology, he regarded this specimen as too precious to use now. He would wait until we were closer to the gene so as to take maximal advantage of the opportunity because, unlike the DNA in the lymphocyte lines, this DNA was not replenishable. Moreover, the DNA obtained from formalin-fixed tissue is variably degraded during the process of tissue fixation. Hence, only a fraction of the DNA fragments might ultimately be intact enough for meaningful analysis.

For a time, I tacitly assumed that no tissue of any kind was available from other deceased affected members of Charles's and Elizabeth's families. But after recalling Charles's visit, I retraced my steps to the various laboratories where autopsies of members of Charles's and Elizabeth's families had been done in the hopes of finding any block of formalin-fixed tissue, even though I knew that the same labs had lost all their frozen tissue as a result of freezer failure years ago. Luckily, however, Dr. Dennis Dickson at Einstein, who had worked with Robert Terry, discovered several small blocks of tissue from two members of Charles's family. For a time, that find seemed the limit of success for this search, but I had overlooked one key figure—Charles. Upon phoning Charles, I learned that he had contacted a number of pathologists over the course of many years and had stored away blocks of formalin-fixed brain tissue from six other affected members of his family. Once again, as a consequence of his foresight, we would have a priceless resource available to use once we came close enough to the gene to justify committing the tissue.

By early 1990 we had in reserve formalin-fixed tissue from eight other individuals in Charles's family, bringing the total of potentially usable DNA samples to eighteen. The numbers could still not approach those in the Venezuelan pedigree, but now we had assembled the largest resource of DNA anywhere in the world from affected individuals within a single family which displayed the classic traits of Alzheimer's disease. I was now fairly certain that we had tracked down all existing sources of DNA from Hannah's affected descendants, unless the abnormal gene had found its way into the Russian branch and these descendants could somehow still be found.

Now and then, in the stillness of a solitary moment, I would think about Hannah and Shlomo's lost descendants perhaps still living somewhere in the vast expanse of Soviet Central Asia. This genetic research was a strange business, encompassing a curious interleaving of endeavors that spanned both time and space.

Inquiries into the genealogical history of Hannah's family had pressed deeper into the past, even as the advancing technologies of molecular genetics raced with promise toward the future. The search for Hannah's descendants in the Soviet Union spanned thousands of miles reaching across to lands east of the Volga, while geneticists estimated the span between the defective gene Hannah left to some of her descendants and the chromosomal sites marked by their probes in the molecular dimensions of nucleotides. Yet through all the efforts of every kind, there was still one single, unifying theme: the quest for an Alzheimer's disease gene, and with the cause of the disease, the conviction that discovery would point to treatment. I needed no further reminder of that objective than the sight of my patients and their families at my Monday afternoon clinic each week.

We never did find Hannah's Soviet descendants. Rather, they found us. Or to be more precise, Hannah's great-grandson, Vasilly Borisevich T., found Bryna. In September 1990, while reading her local newspaper, Bryna's eyes fell upon a small ad that read: "Vasilly Borisevich T. wishes to re-establish contact with his relatives in America." He noted that his Uncle David T. had lived in the area in the early 1930s and requested that anyone with information about the family contact him at a Ukrainian address. As Bryna read on, her eyes fell upon the sentence, "Grandmother was Rosa Shlomovna N." Knowing the Slavic use of patronymics, Bryna recognized the phrase to mean "Rosa daughter of Shlomo N." It was now 150 years since Shlomo had been born in Bobruisk, Byelorussia, and now his name had appeared in a Midwestern American newspaper.

Bryna immediately sent the clipping to me. Yes, of course, I would contact our colleagues in Moscow and have them try to reach Vasilly Borisevitch, and I would dispatch a letter to him immediately. Vasilly had to be the son of Beryl. Borisevich, which means "son of Boris," had to be the Russified name for "son of Beryl." But all this might take time. Perhaps Vasilly was no longer in Russia, after all, Jews were now leaving Russia in the tens of thousands each month for Israel and in the thousands for the United States. I asked Bryna to check with the staff at the newspaper and find out who had placed the notice and when it had been received. Perhaps friends of Vasilly were already living in the United States and could tell us more immediately. Yet no one at the newspaper seemed to know how or when the notice had arrived. Perhaps it had come from a central office of some agency; perhaps it had been sent years ago.

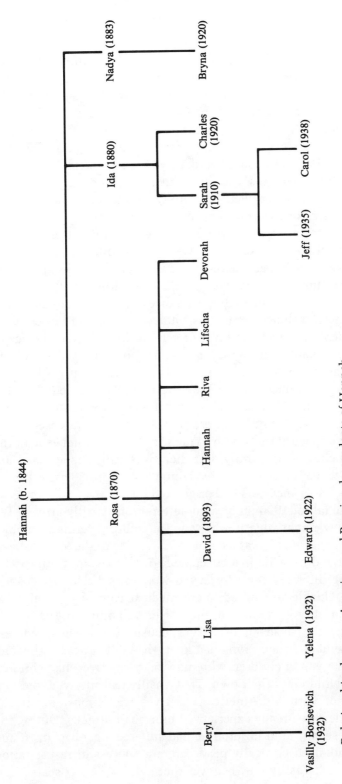

Relationships between American and Russian descendants of Hannah.

I quickly wrote a letter to Vasilly, first giving him some news of his closest American relatives and their addresses, and then explaining the purpose of our inquiry with respect to the Alzheimer's disease gene. Then, realizing that he probably spoke no English and would have to have the letter translated, I asked a Russian colleague to add a brief preface in Russian, "David T. died at age 72 in 1970. His son Edward T., your first cousin, is alive and well and lives with his wife and three daughters at . . ." Thus, Vasilly, if he ever received my letter, would immediately know that his search for close family had been successful.

Within several weeks, Vasilly did receive my letter and contacted Dr. Voskresenskaya in Moscow. He offered to travel to Moscow and give her all the medical information on his family, but his trip was not necessary. His grandmother, Rosa, had lived until 71 without any signs of Alzheimer's disease, nor had any of Rosa's descendants developed dementias. Moreover, most of her children and grandchildren had lived free of symptoms well past the age of risk for Hannah's illness. (Nor had any of Rosa's descendants acquired amyotrophic lateral sclerosis.) My reaction on learning this news was mixed. While I was profoundly relieved that this branch of the family had been spared, I was disappointed that we had been unable to find more cases of Alzheimer's disease, which would have speeded the search to identify the abnormal gene in the other branches of the family. But there was still a hope that Hannah's descendants in the Soviet Union could provide the key to where Hannah's ancestors had originated.

As the fall of 1990 progressed, letters from Vasilly arrived. We learned how remarkable it was that we could have made contact with him at all. Vasilly, as a child of 9, was one of the four children that Beryl and his wife had evacuated to Uzbekistan in July 1941, confirming the story the 87-year-old woman had told Dr. Voskresenskaya. Unable to find adequate medical care in Central Asia, Beryl, then only 42 years old, died of heart failure in 1942. That same year, the husband of Beryl's cousin Lisa—whose address cousin David and then Bryna had kept—had fallen near Stalingrad, leaving two young children who, now 50 years later, had children and grandchildren of their own. That Vasilly had survived to establish contact with his American relatives was almost a miracle. Now 58, he had already experienced four heart attacks. During the last of these, while in hospital he had suffered a cardiac arrest and been pronounced clinically dead before a successful resuscitation was achieved. Vasilly now believes that a divine power spared him so that he could reestablish contact between the branches of his family in the Soviet Union and those in the United States.

Although the Soviet branches had escaped the Alzheimer's gene, they experienced deprivations that were in many ways as tragic as those endured by many of their American kin. Vasilly's father, Beryl, had been a kind, compassionate, and generous man who, when the famine engineered by Stalin struck the Ukraine in the early 1930s, went out of his way to feed the poor and hungry. In 1933 an "invidious scoundrel" denounced Beryl, accusing him of one of the many trumped-up charges that were so common at that time, and Beryl spent years in prison before the outbreak of World War II. The respective husbands of several of Beryl's five sisters had been killed during the war; however, each of the six branches of the family in Beryl's generation had left offspring now reaching into the fifth and sixth generations. Many had died of early heart disease. Whether the risk was caused by the same gene for hyperlipidemia that Charles and others in the United States had inherited, or the result of dietary problems in that part of the Soviet Union, compounded by the great social upheavals and industrial stresses, is still not clear.

As I read more of Vasilly's letters and learned of his great concern for his children and grandchildren, and of his hope that they would be spared the hardships and injustices that had plagued his forebears, I was reminded of the almost identical sentiments of his second cousin, Carol, Jeff's sister, whom he had never known: "I long for the day that my niece and nephews, especially, will be released from the ghosts of their ancestors." It was as if the same spirit had moved both hands and kindled both's letters. It was once again a time to pause, to contemplate, and to hope.

Eventually, Vasilly contacted the other members of his family in the Soviet Union and found that no one knew more of Hannah's origins. This trail was still cold, and so our search for Hannah's origins had finally ended—or so I thought.

Thus, in December 1990, it was quite a surprise when Edward T. received a letter from Israel written entirely in Russian, save for his address. He asked me if I could have it translated, and we discovered that the letter was from Vasilly's first cousin Yelena—and hence also a first cousin of Edward—who had recently immigrated to Israel with her children and grandchildren. Yelena had recently received Edward's address from Vasilly. She included a telephone number, and I asked Edward's permission to phone her.

One afternoon I arranged for Dr. Joseph Tonkonogy to place the call while I stood by. We reached Yelena late one evening in early January 1991. After assuring her that Edward had received her letter, we explained the purpose of our call. There was an excited voice at the other end of the line, a brief silence and then, "Excuse me for pausing, but this is our first contact since 1933!" After further assuring her that her branch of the family did not carry the Alzheimer's gene, we asked her if she knew anything about Hannah's origins. As I heard the word "Litva," my face must have registered a thankful smile because Joseph held the phone nearer to me, but that was her only word in Russian that I understood. "Joseph," I implored, "please ask her if Hannah's ancestors, not just those of Shlomo, had come from Lithuania." Then once again, I heard "Da, da, Litva, Litva!" Later Joseph continued, "She is certain that the ancestors of both Hannah and Shlomo came to Bobruisk from Lithuania."

By now, I had traced the family name of Elizabeth's unaffected great-grandfather to a prominent Lithuanian Jewish family that had lived in the Kaunas (Kavna or Kovno) area from at least the mid-1750s. I also had learned that the family name of Elizabeth's affected great-grandmother had been a common one in Lithuania at that time. If Charles's and Elizabeth's families were related through a bearer of the Alzheimer's disease gene, then perhaps their common remote ancestor had lived in Lithuania, not Byelorussia.

The phone call having ended, I turned to Joseph and, partly out of wishful thinking and partly out of admiration for his vast knowledge of the medical community within the Soviet Union, I challenged him, "Find me the name of a neurologist or psychiatrist in Lithuania who knows about Alzheimer's disease there."

Three evenings later Joseph excitedly phoned me at 11 P.M., "Dan, Dr. Gutman, a young pediatrician, has recently arrived from Kaunas. He distinctly remembers from medical school that presenile dementia—Alzheimer's disease—is well known in Lithuania in both Catholic and, before the Holocaust, in Jewish families. His father is on the medical faculty in Kaunus, and he will forward your inquiry to him." Then Joseph had another surprise: "Dr. Gutman has a friend who has recently arrived in Connecticut from Vilnius. His mother experienced a decline in recent memory in her late forties and was just diagnosed as having Alzheimer's disease at age 51." Her symptoms, as Joseph described them, and their age of onset were virtually identical to the early symptoms of my first patient, Jeff.

For the next seven days, Dr. Gutman tried to reach his father but a Soviet crackdown on the fledgling Lithuanian democracy was now threatened, and he could not get a phone line through to Kaunas. Instead, he sent a letter. Three days later, the anticipated Soviet crackdown occurred, and fourteen Lithuanians in Vilnius were dead. Several days later, the war in the Persian Gulf began, and Scud missiles were falling on Yelena's new homeland.

For Vasilly Borisevitch T., his lifelong objective to reestablish contact between the nine branches of his family had been accomplished. For us, Vasilly, through his contact with Yelena, had directed the search for Hannah's ancestors to the Baltic coasts.

As I examined the new information in terms of what it meant for discovering perhaps the common origin for the Alzheimer's disease gene in Charles's and Elizabeth's families in Lithuania, I recalled Ben's long-held hunch based on the remarkable similarities in the expression of Alzheimer's disease in his family, the British Canadian family, well-documented German families, and several well-studied Swedish families. Ben had hypothesized that perhaps a common gene for Alzheimer's disease had been spread around the Baltic lands—and from there across the North Sea—by travelers, sailors, or invading armies in times long past. Perhaps the Baltic Sea was our Lake Maracaibo.

In principle, Ben's hypothesis would someday be testable. If the genetic defects in these families turned out to be identical, then the various families either shared a common ancestor in whom the mutation arose, or the same mutation arose *de novo* in the several families. If the affected individuals in the various families shared a common affected ancestor, then the sequence of bases of other genes on either side of the Alzheimer gene would in most cases be essentially identical.

By the end of February 1991, I had reached Dr. Aron Gutman at the Kaunus Medical Institute. He conveyed my inquiry to Professor Egidijus Jarzhemskas, chief of neurology and psychiatry in Kaunas. Dr. Jarzhemskas and his younger colleague, Dr. Raimundas Alekna, agreed to undertake the search for families with Alzheimer's disease in Lithuania and the neighboring Baltic lands. As I mailed a detailed letter off to Drs. Jarzhemskas and Alekna, I wondered once again how long it might be before the first gene for Alzheimer's disease would be discovered. Would it be years or might it be tomorrow?

This time it was tomorrow.

The First Abnormal Gene Is Found

By the fall of 1990, evidence for linkage of at least one form of familial Alzheimer's disease to a site on the long arm of chromosome 21, near our originally proposed locus, had firmed up considerably. Our research group, led by Peter Hyslop, joined forces with John Hardy's group in England, Christine Van Broeckhoven's group in Belgium, and several other groups in the United States. We pooled the results on 48 unselected pedigrees with familial Alzheimer's disease and recalculated the lod scores for chromosome 21 markers at and near the sites marked by the original probes. We soon confirmed the conclusion of Allen Roses's group—namely, that most families with forms of Alzheimer's disease beginning after age 65 did not link to markers on chromosome 21. On the other hand, the lod score for linkage to the D21S1/D21S11 markers for families with mean onsets of disease under age 65 now rose to 4.5, with the site of the abnormal gene likely to be within 17 million base-pairs of our closest marker. The odds that there was at least one abnormal gene for Alzheimer's disease in this region of the chromosome were now convincingly greater than 1,000:1.

Four of the pedigrees were likely to be linked to chromosome 21 markers. One of these was the Italian family studied by our group; two others were Belgian families studied by Van Broeck-hoven's team; and the fourth family was a newly discovered British family studied by John Hardy's group. Charles's family and Eliza-beth's family continued to show small positive lod scores at the same D21S1/D21S11 locus, but these lod scores had not increased over the past three years, even though the informativeness of the pedigree had grown with each new case of Alzheimer's disease that we had discovered. As before, we still could not be certain whether the abnormal gene in Charles's or Elizabeth's family resided on chromosome 21; nor was there any compelling evidence against this localization.

The main problems remained technical in nature. Despite years of work, Peter and many workers in other laboratories had not been able to come up with decisive probes for the crucial region of chromosome 21. New probes were always being discovered, but most of these were noninformative, in the sense that the same form of the restriction fragment length polymorphism, or RFLP, was found in both the affected and unaffected parents of our patients, so that these probes were useless as markers for the dis-ease. Moreover, when informative probes were newly discovered, they were so close to our previous probes that they added relatively little new information about the location of the sought-after abnor-mal gene. Inasmuch as Peter could still not be certain whether or not the mutation in Charles's and Elizabeth's family was on chro-mosome 21, he was still reluctant to commit the limited resources of formalin-fixed tissue from deceased affected individuals until we were assuredly closer to the site of the mutation in these families.

Still, the evidence for linkage to markers on chromosome 21 was substantial in four families. The suspected region wherein an abnormal gene lay hidden included a strip of about 15 million base-pairs on either side of the site marked by the D21S1/D21S11 probes. This strip included the recently identified site of the "amy-loid gene," now well known as the amyloid precursor protein (APP) gene. The β-amyloid protein found in the senile plaques and fine cerebral blood vessels was a smaller fragment cleaved away from its normal larger precursor by some as yet not fully understood process. It was the invariable association of amyloid plaques with Alzheimer's disease that had first raised suspicions that a mutation in the amyloid gene might be a cause of Alzheimer's disease. Even so, the MGH group, using traditional linkage methods, had found statistical evidence against a primary abnormality in the APP gene in the Italian family, in the British Canadian family, and in Charles's and Elizabeth's families. Similarly, Van Broeckhoven's group had found that the amyloid gene was not the site of the abnormal mutation in their two large Belgian pedigrees. However, an abnormality in the APP gene had not yet been ruled out in John Hardy's recently discovered British family.

Now a curious set of circumstances and results would soon yield discoveries of great significance. For years neurologists have known of a condition that sometimes occurs sporadically and sometimes on a genetic basis, known as cerebral amyloid angiopathy. In this condition amyloid accumulates in the walls of cerebral blood vessels and leads to brain hemorrhages in adults as early as their fifth and sixth decades of life. One particular hereditary variety of this disease was described in four families living in two coastal villages in the Netherlands. In 1987 Dr. Blas Frangione in New York, working with investigators in Leiden, discovered that the amyloid deposition in the blood vessels of these patients was closely related to the β-amyloid found in both the blood vessels and brains of patient's with Alzheimer's disease.

Frangione's group also discovered that the amyloid deposition was not limited to the walls of the blood vessels, but was found within the substance of the brain as well. In their patient who came to autopsy, the group noted diffuse deposition of amyloid in the cerebral hemispheres, but the researchers did not find the same type of "dense amyloid cores" found in the typical fully developed plaques characteristic of Alzheimer's disease. Moreover, their patient was not demented at the time of death. Frangione and his colleagues proposed the provocative hypothesis that the "deposition of β-protein in brain tissue seems to be related to a spectrum of diseases involving vascular syndromes, progressive dementia, or both." In the conclusion of their paper, the authors went a step further: "Our findings imply that hereditary cerebral hemorrhage with amyloidosis-Dutch is a distinct type of familial Alzheimer's disease with predominant vascular involvement. We propose that this clinicopathological entity be designated familial Alzheimer's disease, vascular type."[1] This conclusion drew limited attention at the time because the Dutch patient was not demented, and dementia has traditionally been the *sine qua non* for considering an entity to fall within the rubric of Alzheimer's disease. (More recent studies by J. Haan and colleagues have shown that most long-term survivors of the Dutch disease eventually become demented.[2] However, it is still not known whether the dementia is caused by multiple cerebral hemorrhages or, conversely, by a degenerative process akin to that in Alzheimer's, or yet another mechanism.) Moreover, the location of the abnormal gene in the Dutch disease was not yet known.

Then, in June 1990, Christine Van Broeckhoven and her group discovered that the disease in the Dutch families linked very closely with the amyloid precursor protein gene now known to be located on chromosome 21. In the very same issue of *Science*, Blas Frangione and his colleagues, in a paper first-authored by Efrat Levy, found the specific mutation in the amyloid gene which caused the Dutch type of hereditary cerebral hemorrhage. They discovered that a "single point mutation"—so named because a "wrong" nucleotide substitutes for the intended nucleotide—in this case, a substitution of the nucleic acid cytosine (C) for guanine (G) at a specific site in the APP gene caused the Dutch disease. This change in a single nucleotide from guanine to cytosine resulted in the substitution of the amino acid glutamine for glutamic acid at a specific site, called position 22, in the amyloid β-protein. This single base change leading to a single amino acid change was apparently all that was necessary to produce the excessive deposition of a slightly altered form of amyloid in the walls of cerebral blood vessels and in brain tissue.

It was against this background that John Hardy's group addressed the question as to whether an abnormality in the amyloid gene might be the cause of the dementia in their British family, which showed substantial linkage to several markers on the long arm of chromosome 21. The family was relatively small, with only six living affected patients from whom Hardy's group could obtain cell lines for DNA analysis. Thus, on the basis of linkage studies alone, they could not determine precisely where the mutation originated. Moreover, as Hardy told me, there still had been no autopsy to confirm the diagnosis of Alzheimer's disease in the British family. In fact, several living members had survived cerebral hemorrhages, raising the concern of Hardy's group that the family might suffer from a variant of the Dutch disease rather than from Alzheimer's.

Then, in July 1991, a victim living near Nottingham was close to death. No less intent than members of Charles's family in furthering research, the family alerted Hardy's group that the patient's death was imminent, perhaps even within hours. Hardy's colleagues Mike Mullan and Fiona Crawford, both on Mullan's motor bike, raced by night over 100 miles toward Nottingham and were able to obtain the patient's brain within two hours after death. After cooling the brain, they raced back to London at breakneck speed until they were stopped by highway police. They quickly explained their scientific objectives, and within an instant, the police volunteered to grant them a high-speed escort to their laboratory in London.

Within hours the neuropathologists established a diagnosis of Alzheimer's disease. Relieved that they were indeed studying familial Alzheimer's disease, Hardy's group now wondered whether a mutation in the amyloid gene could also cause forms of Alzheimer's disease. After all, the amyloid gene had not yet been excluded as a site for the mutation in this family. Hardy's group hit upon a brilliant first strike: they decided to examine directly the sequence of nucleotides in the very same coding region, or exon, of the amyloid gene that harbored the Dutch mutation which led to hereditary cerebral hemorrhage with amyloidosis. On February 21, 1991 Hardy's team at St. Mary's Hospital Medical School in London reported the discovery of a new mutation in this same exon. It was called exon 17, according to its position on an ordered map of the coding regions comprising the gene, but the Alzheimer's mutation was discovered at a different position on exon 17 than that found in the Dutch disease.

The "Hardy mutation" consisted of the nucleic acid cytosine (C) in place of thymine (T) in exon 17, which caused a valine to isoleucine substitution at amino acid position 717 in the amyloid protein. The substitution of a single base was consistently found in all six affected members of his British family, and the normal sequence was found in all the unaffected members. The researchers then obtained similarly consistent results in a smaller American pedigree that Allen Roses's group at Duke Medical Center had previously found to link to chromosome 21.

Hardy's initial report, in a paper first-authored by Alison Goate, and published in *Nature* in February 1991 was cautious.[3] The investigators emphasized that the number of affected individuals with this mutation was small, and that the mutation had not been found in a number of other families with either late-onset or early-onset Alzheimer's disease. Even so, if the British workers were correct in their interpretation of the significance of their discovery, they would for the very first time have obtained strong evidence that a mutation in the APP gene can cause a form of familial Alzheimer's disease. Hardy's group further postulated that the deposition of the β-amyloid peptide might truly be "the central

From left to right: Drs. John Hardy, Fiona Crawford, and Mike Mullan.
(Courtesy of Dr. Hardy.)

event in the pathogenesis of the disorder"—at least in those few
families that harbor this particular mutation.

For the next three to four months, most of the research groups
searching for Alzheimer's disease genes dropped, or at least sharply
curtailed, whatever else they were doing and attempted to determine whether the mutation found by Hardy's group was present in
their pedigrees. Peter Hyslop, who by now had returned home to
Toronto to establish his own laboratory there, and Rudolph Tanzi,
who remained at MGH, worked on testing for possible mutations
in the amyloid gene in other families. By direct sequencing of exon
17, they soon proved that the rare mutation found by the British
group occurred neither in Charles's family nor in the other families
who had been the subjects for the original report linking
Alzheimer's disease to this general region of chromosome 21. Similarly, Gerard Schellenberg's group failed to find the newly described
mutation in their Volga German families. Nor could Van Broeckhoven's group find a mutation in exon 17 of the APP gene in their
two Belgian pedigrees.

Clearly, the mutation described by Hardy's group was a rare
one, but was it the cause of the disease in their two reported families, or was the single base substitution merely a nonpathogenic, or
harmless, mutation with the causal defect lying elsewhere on the

Dr. Gerard Schellenberg. (Courtesy of Dr. Schellenberg.)

Dr. Christine Van Broeckhoven. (Courtesy of Dr. Van Broeckhoven.)

Dr. Allen Roses. (Courtesy of Dr. Roses.)

chromosome? In the early spring of 1991, that question could still not be settled; yet within a few more months, definitive evidence that the newly described mutation did actually cause the disease rapidly fell into place. The same mutation was found in four more pedigrees with a familial form of Alzheimer's disease coming from diverse racial backgrounds—one family from France and three from Japan. Moreover, DNA from hundreds of normal individuals all over the world had now been studied with respect to the mutation in exon 17 of the amyloid gene, and no one had reported the mutation occurring in the general population. Thus, by the late spring of 1991, Hardy's group had convincing evidence that the single base substitution in exon 17 was indeed a cause of a form of familial Alzheimer's disease, though a rare one.

Hardy's group now turned their attention to the neuropathological findings found in the six families that showed their mutation. In the British family, only the one autopsied case was available for study. The brain showed both the senile plaques and neurofibrillary tangles characteristic of Alzheimer's disease, as well as amyloid depositions in the small cerebral vessels consistent with a mild amyloid angiopathy which so often accompanies Alzheimer's disease. However, several other surviving members of the British family had symptomatic "amyloid angiopathy," with cerebral hemorrhages reminiscent of the symptoms found in the Dutch form of cerebral amyloidosis. In the American family which had the same mutation, the clinical picture was also consistent with typical Alzheimer's disease. Here two brain autopsies showed numerous amyloid plaques and neurofibrillary tangles, but in this case no amyloid angiopathy was found. The finding of both plaques and tangles in these cases without an accompanying amyloid angiopathy was especially important: it proved that the same abnormal gene can produce brain pathology typical of Alzheimer's disease even in the absence of the deposition of amyloid in cerebral blood vessels. In two of the Japanese cases that came to autopsy, the researchers found standard Alzheimer's disease pathology and amyloid angiopathy within vessels in various parts of the brain.

As a result of these studies, Hardy's group realized that there were fine differences in the expression of the disease both within and between the families that had the same mutation. Modifying factors as yet not understood, which may vary both within and between families, may produce a varying spectrum of pathological findings in different individuals.

Thus, Hardy's team convincingly established the molecular

basis for the genetic defect in at least one form of familial Alzheimer's disease. (In recognition of the significance of these results, the paper of the St. Mary's group became the most cited report in the biomedical literature for the year of 1991, as reported in *Science Watch*[4] in January 1992.) Hardy's group also demonstrated that the classification of various forms of familial Alzheimer's disease must henceforth be based on a classification at the molecular level rather than on a selected clustering of "phenotypic," or "expressed," features. Thus, the work of the British team confirmed and extended the idea of Frangione's group that there may be an extended spectrum of expression for a single genetic disorder. In the cases reported by Hardy's group, some showed amyloid plaques and tangles in the brain matter itself, some showed only amyloid deposition in cerebral vessels, and still others showed both types of findings. Even this single gene mutation had a variety of expressions.

Many geneticists were stunned that two phenotypically distinguishable diseases—cerebral amyloid angiopathy and one form of familial Alzheimer's disease—resulted from two distinct mutations that were not only within a common gene but even within the *same* exon of that gene. Their surprise at this result would soon give way to even greater astonishment. For as other research groups turned their attention to determining whether the "Hardy mutation" was the cause of Alzheimer's disease in their own patients, two additional mutations were discovered. Both were in the very same "codon," the same ordered triplet of nucleotides encoding a specific amino acid, in this case numbered 717. This sequence normally encodes for the amino acid valine. In early October 1991, the group of Jill Murrell and Merrill Benson at the Indiana University School of Medicine reported in *Science* that another single amino acid substitution (phenylalanine for valine)—a result of a mutation in the same codon—also produces familial Alzheimer's disease.[5] Later that month reporting in *Nature*, the St. Mary's group, this time in a paper first-authored by Marie-Christine Chartier-Harlin, revealed a third single-base substitution as a consequence of yet another mutation in codon 717; this mutation produced a glycine for valine substitution, and this too could lead to familial Alzheimer's disease.[6]

Any of three single-base substitutions in codon 717 of the amyloid precursor protein gene could produce familial Alzheimer's disease: if amino acids isoleucine, phenylalanine, or glycine substituted for valine, the consequences could lead to familial Alzheimer's disease.

No one knew precisely how these three distinct mutations in exon 17 of the amyloid gene could cause the pathological changes of Alzheimer's disease to develop, nor do we know today. However, a number of reasonable hypotheses were aired. Most of these hypotheses began with the extremely important observations of two groups—the one of S. S. Sisodia, E. H. Koo, and Donald Price at Johns Hopkins, Konrad Beyreuther now at Heidelberg, and Axel Unterbeck now at New Haven, and the other of Fred Esch and associates from San Francisco—who showed that the breakdown of the amyloid precursor protein normally, or at least most always, occurs by cleavage of this molecule *within* the β-amyloid region. Hence, no intact β-amyloid fragments of the type found in Alzheimer's disease can be generated during the breakdown of the amyloid precursor protein along this metabolic pathway.

However, in normal aging to a lesser extent and in Alzheimer's disease to a greater extent, some shift might take place in the metabolism of the amyloid precursor protein. Accordingly, this protein must be split more frequently along some alternative metabolic pathway, which then results in the production of numerous intact β-amyloid fragments. Many workers suspected that the split β-amyloid fragments may themselves be toxic to brain cells and ultimately cause more damage to the nerve cell membranes than was caused by the actual abnormal cleavage of the precursor molecule.

Other scientists, however, remained unconvinced that the β-amyloid fragment could cause the pathological changes. For it was still possible that the defects which lead to nerve cell death and Alzheimer's disease resulted from abnormalities in the cleaved larger fraction of the amyloid precursor molecule, the part that remains within the membrane rather than the β-amyloid fragment that has found its way to the extracellular spaces outside the neuron which may cause toxicity. Presumably, the mutations in exon 17 could somehow lead to an excessive abnormal cleavage of the amyloid precursor protein and a concurrent excessive deposition of a β-amyloid fragment. Indeed, even today, we still do not know whether the amyloid deposition is a primary cause or a consequence of a more fundamental process. Moreover, whether the β-amyloid fragments could reach the brain and cerebral blood vessels from the bloodstream, or by degradation of the amyloid precursor protein from cells in both the brain and the walls of blood vessels, or by both mechanisms, also remains unresolved. However, several working models were proposed at the time to explain how the altered metabolism of the amyloid precursor protein might produce the cascade of pathological events that ultimately leads to cell damage and to dementia.[7]

Several groups of workers attempted to determine whether the β-amyloid fragment is itself the toxic agent that causes cell death and the pathological changes seen in Alzheimer's disease. Bruce Yankner's group, comprised of researchers from both the Children's Hospital and MGH in Boston, extending work begun with Rachel Neve, reported evidence that the β-amyloid fragment is toxic to certain brain cells in tissue culture. These findings were confirmed by several research teams.[8] Yankner's group also believed that toxic changes occur within the brain of normal animals following the injection of β-amyloid directly into their brains; moreover, they maintained that these changes were highly specific and closely matched those found in Alzheimer's disease, and yet were quite different from lesions produced by certain other neurotoxic agents. Some research groups, but not others, confirmed finding some neuronal degeneration in animal brains following β-amyloid injections. However, in the brain injection studies, the results were less consistent and highly controversial—in fact, so controversial that Paul Coleman of the University of Rochester devoted the entire September 1992 issue of the *Neurobiology of Aging* to papers taking one side or the other on the question of whether β-amyloid is toxic to brain cells.[9]

But, if Yankner's ideas were correct, we would soon see attempts to treat at least part of the symptomatology of some forms of Alzheimer's disease with chemical agents. These agents could either slow the excessively rapid breakdown of the amyloid precursor protein or interfere with the access of these presumably "toxic" β-amyloid fragments to normal nerve cell membranes. In the fall of 1992, it was much too early to judge whether the hypothesized toxicity of β-amyloid on certain brain cells, even if confirmed, would be a primary or an intermediary step in the causation of Alzheimer's disease. Or would it possibly be a primary cause in a minority of cases of familial disease and a secondary, contributing cause in the majority of other familial cases, as well as in the late-onset sporadic cases? Even so, studies such as Yankner's showed us how quickly we might be able to go from the level of a proved genetic deficit to a mechanism that kills brain cells and that might be amenable to the development of selective pharmacologic therapy, perhaps even without the direct need to "repair" a defective gene. In fact,

Yankner's group reported in October 1990 that "treatment" by the injection of a peptide, called substance P, either directly into the brain or into the bloodstream could prevent the toxic response of brain cells to β-amyloid. This finding has yet to be replicated by other groups.

Although the mutations discovered by Hardy's group remained of major significance, their relative rarity provided all the more incentive for geneticists to press on to discover the abnormal gene in Hannah's family and in the vast majority of other families affected with a familial form of the disease not caused by a primary defect in the amyloid gene. In these families—and perhaps in cases of late-onset Alzheimer's disease as well—some other primary molecular abnormality must set off the cascade of changes that leads eventually to cell death and excessive amyloid deposition. In these cases it may be just as, or more, important to discover and treat the primary abnormality as to attack only the later stage of amyloid deposition. Thus, first causes were still to be sought, and as speedily as possible.

For a brief time beginning with a publication by Shigeki Kuwabata, Gerald Higgins, and Jon Gordon in *Nature* on December 12, 1991, the central problem of a major cause for Alzheimer's disease appeared to have been solved.[10] These workers produced a "transgenic" mouse in which human DNA coding for a fragment of the amyloid precursor protein, which Yankner and Neve believed to be toxic to neurons in tissue culture, was inserted into the genome of reproducing mice. A neuron-specific gene "promoter" was tied onto the injected DNA so that the human-derived amyloid fragment would be selectively overexpressed in the brain cells of the mice. The results reported were spectacular: "These transgenic mice demonstrate that a single disturbance, overproduction of the carboxy terminal portion of APP, can lead to all of the major neuropathological features observed in AD."

The report seemed to clinch the case that a neurally overexpressed amyloid fragment was sufficient cause to produce all the pathological features of Alzheimer's disease. However, within two months, this transgene story began to unravel. Several accounts of what may have gone amiss have now been published. John Rennie's investigative report,[11] published in the June 1992 issue of *Scientific American*, said the following:

> Several neuropathologists who have looked at slides provided by Higgins say they have found them far less persuasive than the images

published in the December paper. Selkoe, for example, recalls that he met with Higgins this past January and was distressed by the condition of the slide he was shown. All the lesions appeared to be concentrated within one fragmented area of the brain tissue. Selkoe thought the lesioned area could be interpreted as a piece of a diseased brain, possibly from a human patient, that was merely adjacent to normal mouse tissue.

Donald L. Price and Larry C. Walker of the Johns Hopkins University School of Medicine, who also examined Higgins's slides, concured with Selkoe's interpretation. The National Institutes of Health launched an inquiry into the matter and would try to determine whether the reported findings were genuine, a result of an accidental mixup of tissues, or a consequence of scientific misconduct.

Subsequently, Kawabata, Higgins, and Gordon were unable to reproduce their neuropathological findings in other transgenic mice and, in March 1992, retracted their original paper. Thus, at present, there is no definitive evidence based on transgenic mice studies in favor of the hypothesis that excess production of the β-amyloid fragment is a sufficient cause of Alzheimer's disease.

New Sightings

The first mutation in the amyloid precursor protein gene reported by John Hardy's group in February 1991, the two other mutations in the same codon, and further mutations in the APP gene discovered in 1992[1] together accounted for only about 4 percent of known familial Alzheimer's cases.[2] But the very fact that these mutations were sufficient to cause so many of the clinical and pathological findings associated with Alzheimer's disease made these discoveries central for any future understanding of the pathogenesis of at least one form of the disease.

I found Hardy's original discovery both tantalizing and ironic. The pathological findings in the affected members of Charles's family had many features in common with those of Hardy's English family, including the rather significant deposition of amyloid in the fine cerebral blood vessels and, notably, the markedly increased incidence of thyroid disease in the affected members of both families. Yet the mutation found in this English family simply did not occur in any of the affected members of Charles's family or the large English Canadian family under study by the MGH group. I found it puzzling that Peter Hyslop's original linkage of a form of familial Alzheimer's disease to the D21S1/S11 marker relatively close to the amyloid gene had pointed the way for the success of the British group, yet the very same group of families who led us to this region of chromosome 21 failed to show any mutations in the APP.

We still did not know whether the genetic defects in our original four families involved a nearby gene that produced its deleterious effect through some still unknown regulatory mechanism affecting the expression of the amyloid gene, or whether there were totally separate genetic mechanisms—perhaps abnormal genes on other chromosomes not primarily involving amyloid at all—that produced virtually identical clinical and pathological pictures. Thus, it was a time for us to take a hard look at both the difficulties and opportunities that lay ahead and, if possible, to plot a surer course. We still thought that the odds remained reasonably good that the genetic defect in the Calabrian family was on chromosome 21, not far from the amyloid gene. However, we would need to reconsider the prospects of finding the genetic defect for Alzheimer's in Charles's family and in the British Canadian family on chromosome 21. All the while, the various research groups looked for advances that might speed the resolution of the chromosome 21 issue. All remained ready to leap upon new developments that would open the way for rapidly testing candidates on other chromosomes.

As 1992 began, six teams remained committed to the search—John Hardy's group still in London but soon to move to Tampa, Christine van Broeckhoven's in Antwerp, Allen Roses's at Duke University, Gerard Schellenberg's in Seattle, Rudolph Tanzi's at the MGH, and Peter Hyslop's team, with whom I still collaborated, now established in Toronto.

Even before the first discovery of a mutation by Hardy's team, interest by all groups in devising an improved map of markers for the Alzheimer's disease region of chromosome 21 had remained high. In one of the few but notable attempts at cooperation between all competing groups seeking genes on chromosome 21, a "chromosome 21 club" had developed under the coordination of David Patterson in Denver, Colorado. Patterson provided an invaluable service to all interested workers by determining the position of new markers for chromosome 21 with respect to existing markers and making the information freely available. Despite all these efforts, in early 1992 there were still not enough "polymorphic" informative markers available in the regions around the D21S1/S11 markers and the amyloid gene to settle the issue as to whether the abnormal gene in Charles's family and the British Canadian family could still be on chromosome 21. Moreover, the evidence from the Seattle group remained compelling that the abnormal gene in the Volga German families did not reside on chromosome 21.

Thus, by mid-1990 several groups around the world had begun scanning the remainder of the genome to see if major genetic defects for familial Alzheimer's disease might reside on other chromosomes. The work was tedious and laborious, but it had to be done. In the absence of further markers that could resolve the question on chromosome 21, the best course, at least with respect to those families that did not link unequivocally to chromosome 21, was to test for linkage of their forms of Alzheimer's disease to other chromosomes.

By the early summer of 1991, Allen Roses and his colleagues at Duke University published a study showing suggestive, but not definitive, evidence that a gene causing susceptibility to late-onset Alzheimer's disease may reside on chromosome 19.[3] The significance of such a result, if confirmed by other groups, would be profound. However, the technical problems in working with late-onset Alzheimer's disease remained formidable, in that Roses's results had been obtained by pooling data on a very large number of small families rather than by study of a small number of large families. We still have no knowledge of a very large family with late-onset Alzheimer's disease that could be used for study, and in grouping many small families with late-onset disease there is always the risk, indeed almost a certainty, that we may be dealing with several genetic risk factors at different sites, and perhaps with multiple environmental factors as well. Even so, the new work on chromosome 19 offered promise, though any results would probably take much longer to sort out. However, as the history of this quest had taught us, this situation could change rapidly. "Candidate genes" could now be tested within months of their discovery in small families, as we saw when John Hardy's group tested the amyloid precursor protein gene.

Thus with every passing year, the vast mass of information on biochemical abnormalities in Alzheimer's disease that had accumulated took on ever greater relevance. For now, once a candidate gene was discovered, geneticists could determine by molecular methods whether a given abnormality is the direct genetic cause of a disease or simply a secondary consequence. The method of seeking "candidate genes" hazards chance, but could lead to rapid discovery, as witnessed in the work of Frangione's and John Hardy's groups.

But, if no likely candidate genes become available for study, the molecular geneticist must rely solely on linkage studies using the family method and DNA probes. Fortunately, another major breakthrough in the use of DNA probes was made in 1989, about the time that geneticists were becoming acutely aware that there were still no useful RFLP probes for numerous regions of the genome.

Working at the Marshfield Research Foundation in Marshfield, Wisconsin, James Weber and Paula May discovered another type of human DNA marker that could become as useful as—and in special cases more useful than—the restriction fragment length polymorphisms (RFLPs) which had revolutionized molecular genetics scarcely a decade before. Their novel idea was based on the discovery that interspersed blocks of DNA are scattered throughout the human genome and contain "repeats" of the nucleotides cytosine (C) and adenine (A). These two nucleotides, or "dinucleotide repeats," also called "short tandem repeats," or STRs, recur at very specific chromosomal sites, with variable numbers of repeats in different unrelated individuals. Thus, the length of these repeats, which reflects the number of repeating cytosine-adenine segments, can be the basis for an inherited polymorphism, where the number of repeating segments might range from 6 to 30 or more. As the 1990s progress, more and more markers based on the dinucleotide repeat technology will certainly become available. And in time, because of the inherently great variability of these repeats within different individuals, the dinucleotide repeats will likely supersede the RFLPs for mapping out certain regions of the human genome.

As we look at the scene in molecular genetics today and try to guess at the future of gene quests, it is difficult to imagine that the vast Human Genome Project will not play a role in the search for some of the remaining genetic defects in Alzheimer's disease. The project has become a highly visible, billion dollar national and international venture.[4] In the United States, it is heavily funded by both the National Institutes of Health and the Department of Energy.

The project's current goals are to provide a set of linkage markers for the entire human genome, to establish a physical map for the genome as well, then to sequence all of the DNA for the coding regions of human genes, and perhaps still later to sequence all 3 billion base-pairs of the human genome.[5] Extensive debate continues as to what its major objectives should be and how to best reach them. Eventually, the project will establish an immensely valuable reference library that should accelerate the search for most genes that have not yet been found by direct attack. Even so, the genome project itself became one of the most controversial topics in the biological sciences in the late 1980s and remains so during the early 1990s. Few biological scientists doubt the importance of obtaining the sequencing information. As James Watson, one of the co-discovers of DNA who became the director of the project, noted:

> A more important set of instruction books will never be found by human beings. When finally interpreted, the genetic messages encoded within our DNA molecules will provide the ultimate answers to the chemical underpinnings of human existence. They will not only help us understand how we function as healthy human beings, but will also explain, at the chemical level, the role of genetic factors in a multitude of diseases, such as cancer, Alzheimer's disease, and schizophrenia, that diminish the individual lives of so many millions of people.[6]

Although few researchers doubt the importance of ultimately sequencing the entire genome, intense controversy has arisen over whether billions of dollars should be targeted for a mapping project, thereby diverting desperately needed funds from perhaps more important research initiated by independent individual investigators. Everyone is aware that this project, gigantic in scope, offers great promise in the long term for aiding the fight against numerous devastating diseases that afflict humankind. However, many honest and bitter differences of opinion exist as to whether in the short run the project will better serve the quest for any given disease gene. Either way, I suspect that several more abnormal genes for familial Alzheimer's disease, on chromosomes other than chromosome 21, will probably be discovered by decade's end, especially if well-organized collaborative efforts on the world's most genetically informative families can be developed and the Human Genome Project's target of a set of closely spaced linkage markers over the entire genome remains on schedule. Moreover, the quest for slowing or preventing the disease process, in contradistinction to the myriad attempts to modify the symptoms associated with the degenerative process, has already begun with the discovery of the first abnormal gene.

Techniques are already available, or falling rapidly into place, which will make the discovery of the remaining Alzheimer's genes virtually inevitable in time, and an attack on the consequences of these genetic deficits ever more hopeful. How fast these successes come may depend less on existing knowledge and techniques than on less predictable factors, such as the willingness of intensely competitive groups of molecular geneticists around the world to pool their resources to speed these quests. In the final analysis the human factor and the willingness or unwillingness of the different groups to cooperate with each other may well set the pace of discovery.

The modern age of molecular genetics, and with it the promise to treat many devastating human illnesses, offers great hope, but the scientific community remains equally aware of potential perils that must be avoided. When James Watson agreed to head the Human Genome Project, his concern for avoiding these perils was as much in his mind as were the vast opportunities that lay ahead. As Watson noted,

> I made clear my concern for the ethical and social implications raised by the ever-increasing knowledge of human genes and of the genetic diseases that result from variations in our genetic messages. On the one hand, this knowledge undoubtedly will lead to a much deeper understanding of many of the worst diseases that plague human existence. Thus, there are strong ethical reasons to obtain this genetic knowledge as fast as possible and with all our might. On the other hand, the knowledge that some of us as individuals have inherited disease-causing genes is certain to bring unwanted grief unless appropriate therapies are developed. So it is imperative that we begin to educate our nation's people on the genetic options that they as individuals may have to choose among.[7]

And as Watson continued,

> We must work to ensure that society learns to use the information only in beneficial ways, and if necessary, pass laws at both the federal and state levels to prevent invasions of privacy of an individual's genetic background by either employers, insurers, or government agencies, and to prevent discrimination on genetic grounds. If we fail to act now, we might witness unwanted and unnecessary abuses that eventually will create a strong popular backlash against the human genetics community. We have only to look at how the Nazis used leading members of the German human genetics and psychiatry communities to justify their genocide programs, first against the

mentally ill and then the Jews and the Gypsies. We need no more vivid reminders that science in the wrong hands can do incalculable harm.[8]

As a consequence of Watson's efforts, the Human Genome Project is devoting at least 3 percent of its funding to explore the ethical and social implications of genetic research. Such funding is only the start. The ethical and social concerns remain a responsibility for all of us involved in research.

As scientists, physicians, and citizens alike, we are well aware that our concern for the ethical and social implications of research is not limited to genetics. For example, the same chemical industry that gave Alzheimer and Nissl the stains to study the human brain gave Ehrlich the means to develop a rational basis for effective pharmacological treatment of infectious diseases later produced Zyklon B and still produces precursors for poison gas. Watson's reminders must never be forgotten.

Mutual Aid as a Factor in Evolution and in Genetic Research

Many times during my involvement in the search for the Alzheimer's disease genes, I have vividly recalled the day when my father introduced me to the theory of evolution. He explained the classical Darwinian view of the struggle for existence and the survival of the fittest, as many of Darwin's most strident followers such as Thomas Huxley had interpreted the theory. But then my father added, "There is another side of the story, and naturalists such as Peter Kropotkin proposed that mutual aid and interspecies cooperation are just as important as factors for the survival of a species." In due course I read the works of both Darwin and Kropotkin, but I could never have imagined then that I would someday observe at firsthand the most extreme forms of these two opposing yet complementary evolutionary stratagems—the one as practiced by my fiercely competing colleagues in molecular genetics, and the other by my cooperating patients and their families.

Peter Alekseevich Kropotkin, the famous geographer, geologist, sociologist, and theoretician of anarchism, was born in 1842. In his mid-twenties, fresh after reading Darwin's then new work, Kropotkin explored regions of Eastern Siberia and Northern Manchuria, in those harsh lands where animals had to struggle against the extreme severity of the climate. Kropotkin expected to find evidence of a bitter struggle for existence between animals as most of Darwin's interpreters had predicted. On the contrary, he noted, "in all these scenes of animal life which passed before my eyes, I saw Mutual Aid and Mutual Support carried on to such an extent which made me suspect in it a feature of the greatest importance for the maintenance of life, the preservation of each species and its further evolution."[1]

These insights lay dormant in the young Kropotkin until January 1880, when he heard a lecture by Karl Fedorovich Kessler, a well-known zoologist and dean of the University of St. Petersburg, who suggested that besides "the law of mutual struggle there is in nature the law of mutual aid, which, for the success of the struggle for life, and especially for the progressive evolution of the species, is far more important than the law of mutual contest."[2] Upon rereading Darwin's *The Descent of Man*, Kropotkin realized that Darwin himself had pointed out how in numerous animal societies, "the struggle between separate individuals for the means of existence disappears and the *struggle* is replaced by cooperation."[3] Darwin even intimated that the fittest need not be the physically strongest nor the most cunning but those, to use Kropotkin's words, "who learned to combine so as mutually to support each other, strong and weak alike, for the welfare of the community." For Kropotkin, "those animals which require habits of mutual aid are undoubtedly the fittest." Kropotkin in his remarkable book *Mutual Aid, a Factor of Evolution* went on to document examples of such aid in all animal society, from ants and other invertebrates, up through the animal kingdom, and finally to humans.

For my part, I have never seen better examples of such mutual aid and support than those shown by Charles's family and by so many other families, also victims of Alzheimer's disease, with whom I have had the privilege to work in recent years. At the start, as Charles explained it to me, "We tried to save ourselves."

Charles will accept no praise for all that he has done; "I believe that anyone in my situation would have done exactly as I did." I do not know if Charles is expressing his characteristic modesty or genuinely believes, in some deterministic sense, that anyone constructed exactly as he was and facing the same situation could not have acted differently. Assuredly, even if Charles's initial objective had been to try to save himself and his kin, he exposed himself, body and soul, to numerous additional stresses and anxieties in the process. Some in his situation have chosen denial or distraction as mechanisms for avoiding the pain and anxiety attendant upon being genetically at risk for one of humankind's cruelest genetic diseases. Most, in the early days, would have considered the quest which Charles began as simply futile, and yet he had the courage to *begin* that quest and the determination to stay with it. Whether Charles's example represents a freely willed course of action or a biologically determined expression of programmed biological behavior, I cannot venture to know. Either way, a species that has evolved a Charles has much to commend it and inspire it in its darker moments.

As the struggle in Charles's family progressed, I believe that I saw an evolution from a restrictive self-interest to a broader concern for all those similarly at risk. Charles's family and all the other at-risk families soon realized, and they probably understood it intuitively from the start, that to save themselves they had to enlist the help of many others, and also be ready to help those more distantly related to them.

In many respects, the ideas of Kropotkin, who recognized the survival value of mutual aid, foreshadowed the efforts of modern-day sociobiologists, who have striven to discover the biological basis for the development of such intraspecies cooperation and to discern the mechanisms by which mutual aid enhances the survival of the gene pool of its practitioners. Edward O. Wilson, a founder of sociobiology, following upon the earlier work of W. D. Hamilton and Robert Trivers, considered the origin of altruism as the central theoretical problem of sociobiology. In common usage, altruism is taken to mean the principal or practice of unselfish concern or devotion to the welfare of others. However, in the field of animal behavior, altruism has come to specify behavior of a person (or animal) that "increases the fitness of another at the expense of his own fitness."[4] Reciprocal altruism—a term first offered by Trivers—then becomes, to quote Wilson, "trading altruistic acts by individuals at different times."[5] Wilson asks, How can altruism which, on occasion, can reduce the personal fitness of an individual possibly evolve by natural selection? The answer he offers, concurring with Hamilton, is *kinship*: "if the genes causing the altruism are shared by two organisms because of common descent, and if the altruistic acts increase the joint contribution of these genes to the next generation, the propensity for altruism will spread through the gene pool."[6]

Thus, to some modern sociobiologists, altruism is not a selfless act in any strict sense of the word, but rather an effective strategy that has evolved within the animal kingdom that well serves the perpetuation of others of like kind. Whether the sociobiologist's explanation for the origin of altruism and reciprocal altruism is complete and correct, I do not know. However, even if the initial impulses toward altruism sprang from raw instinct or from an enlightened self-interest, the finished product—like a fine sculpture chiseled from a block of marble—has far transcended its humble origin. In this same sense, our cultural values and heritages, although offsprings of our genes, in time establish supremacy over our raw genetic endowment.

I also believe that the initial resolve of those of us close to victims of Alzheimer's disease, whether or not we were initially motivated by principles of reciprocal altruism, extended beyond our own limited circle and evoked support, sympathy, and help from an ever-enlarging community of our fellow humankind. Perhaps the common decency and humanity of our species is also of sociobiological value, an appreciation of both the gifts of genetic diversity and the universality of the human condition. This appreciation is recognized by geneticists, ethicists, and poets alike, and is perhaps no better expressed than by John Donne:

No man is an Iland, intire of it selfe;
every man is a peece of the Continent, a part of the maine;
if a Clod bee washed away by the Sea, Europe is the lesse,
as well as if a Promontorie were, as well as if a Mannor
of thy friends or of thine owne were;

Yet, for the most part, my colleagues in molecular genetics were more subject to Huxley's view of Darwinism and in as harsh a form as any of Darwin's contemporaries would ever have imagined. Virtually all of my scientific colleagues entered their fields out of a genuine thirst for knowledge and an intense desire to help humankind. Their motivation to go into science, as was mine, was partly driven by the search for excellence, partly by the quest for discovery, and in no small measure by the desire to be the first to appreciate the significance of a new fact of nature. The objectives of the scientist in this respect are no different from those of the athlete to excel, the musician to create a new opus, or the explorer to find new lands. We share vicariously in their triumphs when we grasp the significance of a new discovery, anticipate, and recreate in our own minds the sounds of a favorite composition or gaze upon a fresh landscape.

Along the way, our scientists have had to excel at every step, for the opportunity to pursue such lofty goals is, in fact, reserved for only the exceedingly capable and precious few who survive the rigors of the educational process and successfully compete for research funds to carry out their work. The pressures and the competition become especially intense when the objective is to discover a gene for an important disease.

The competition is heightened in molecular genetics as in perhaps no other area of biology. The discovery and sequencing of a gene is a singular event. The discoverer will achieve recognition in the history of the field. Tens of thousands of discoveries by others may have led the way, but only one set of workers will usually be recognized for the final step in the quest. And the sequencing of a gene is not simply an end in itself; rather, it is simultaneously the beginning of years of serious work in the cell biology of how different transcripts of the gene are expressed in different cells and tissues, and the determination of the functional role of the various expressed proteins. There is little credit for second place in molecular genetics. The drive to be first is heightened by numerous other factors, with the Nobel Prize often looming in the background.

Financial rewards are larger now than perhaps at any time in history. The discovery of a gene, and with it tests for the presymptomatic diagnosis of a disease and perhaps its eventual treatment, can lead to great financial rewards for research scientists, the institutions that sponsor their work, and biotechnology companies. In the process, considerations that are normal and appropriate in the business world can erode the traditional ethical standards of the university-based academic research scientist. At times, we first learn of a discovery in molecular genetics only after the scientist has filed his patent application, even if the scientist is one with whom we are working closely.

If perceived rewards are high, so too are the risks. Research funding is hard to come by, even for the best of causes. Honest errors or premature judgments by one group of scientists can seriously affect the chance of funding for others. For a time, when it was thought that one group already had the cystic fibrosis gene in hand, many new groups seeking support for the same task could no longer obtain funding. Leslie Roberts writing in *Science* documented many of the intense personal struggles between scientists that have accompanied several successful gene hunts.[7]

Moreover, the field of molecular genetics is largely an endeavor fueled by technology. The availability of so many powerful techniques makes an attempt to solve a problem open to numerous groups all over the world, and almost always within the same time frame. As we noted earlier, when the amyloid precursor protein gene was discovered, four groups independently published their work within a three-week period in February 1987. The literature contains fewer references to the groups that published similar findings only a month later. A scientist, especially a lone young scientist, can find his or her whole career threatened if five to ten years of pioneer work is suddenly picked up at an opportune time and exploited by a large team in another lab.

With such pressures of competition, the practice of sharing scarce and unique resources, such as cell lines and DNA probes, is not yet the norm. In fact in May 1991, when two research groups successfully collaborated to link one form of amyotrophic lateral sclerosis, or Lou Gehrig's disease, to chromosome 21, Michael Conneally, a distinguished expert on genetic linkage mapping, commented in an editorial in the *New England Journal of Medicine*: "Thus, this study is one of the few instances in the field of gene mapping in which the investigators themselves initiated collaborations pooling all families, probes, and resulting data for both groups. It is hoped that the organization of this study may be adopted by other investigators who are attempting to map rare genetic disorders."[8]

Allen Roses and Conneally in a letter to *Science* on April 5, 1991 reminded the scientific community that an Alzheimer's Disease Research Center National Cell Bank has been in operation for prospective donations since September 1989 so that it could subsequently provide cell lines "to investigation of the genetic defect that is associated with Alzheimer's disease."[9] As of late May 1992, Roses regretfully acknowledged to me that this cell bank and another in Camden, New Jersey, have yet to fulfill their objectives with respect to donations of cell lines.

Many workers in the field believe that all products developed by investigators whose work is supported by government funds should be available to all other scientists immediately upon publication of the results. Indeed, a directive from the National Institutes of Health announced on March 18, 1988 recommends this policy. Yet there are times when the immediate implementation of such a policy can threaten the career of a young scientist who has worked for years developing a cell line or completing the sequencing of an important gene. Must the scientist share the fruits of his labors immediately because the next, perhaps culminating, experiment can be done far more rapidly by a larger established laboratory that has greater financial resources and staff than the young discoverer? Writing in *Science*, Eliot Marshall has thoughtfully addressed the intricate complexities of many of these difficult issues.[10]

The openness of ideas and the cooperation between scientists that Niels Bohr tried to introduce into biology through Max Delbrück did not become the standard in molecular genetics, though there are many notable and admirable exceptions. Nor did such openness long endure in Bohr's own beloved physics. With the advent of the possible development of atomic weapons, such openness vanished with the secrecy in nuclear research imposed by the various governments before, during, and after World War II.

Selective collaborations among competing groups of molecular geneticists have sprung up, but usually only *after* one or another group has already staked out a credible claim for linking a disease to a general region of a particular chromosome. At present, at least six major research groups are independently scanning the human genome, seeking hints for the localization of those abnormal genes for Alzheimer's disease that do not link to chromosome 21. The number of large informative families available for such studies remains only a handful. Yet to date no worldwide cooperative effort has been made to scan systematically the entire genome in these most informative families to find the abnormal gene(s), although smaller cooperative ventures such as between the Toronto and Duke groups are proceeding nicely. Given human factors such as personality conflicts and professional rivalries, which are sometimes intensely bitter, and the technical and logistic problems in large-scale cooperative ventures, as well as the possibilities of a diagnostic error leading to worldwide confusion in a gene search, it is an open question whether individual endeavors will yield new results faster than a cooperative approach.

Whether or not research groups wish to collaborate among themselves, the sharing of cell lines is an issue apart. In my view, these lines, freely given by our patients and their families and the corresponding information about the clinical status of each donor, are more analogous to raw materials than to products in development or finished goods. I accept that a scientist need not divulge to competitors as yet unpublished findings or share advances in technique that may be the results of years of arduous labor, struggle, and ingenuity. But the basic materials are another matter; they often comprise the basis for the only reasonable starting point for a particular research problem. Hoarding these raw materials only thwarts those competitive efforts that can accelerate progress. Thus, I can discern no ethical justification for any research group to withhold from another the cell lines of any stricken family with any genetic disease after some reasonable time. The provisos are that the confidentiality of all patients be maintained, that the requesting group have the funds to cover the expenses for preparing and distributing the cell lines, and that the providing group be given sufficient lead time to prepare the samples without interrupting its own research. (Of course, the last qualification becomes moot if the cell lines have already been "banked" with one of the several existing cell-line repositories.)

The qualification of "a reasonable time" is probably needed for practical reasons, because investigators must expend considerable time, energy, and financial resources to research a geneology and clinically work up an informative pedigree. Without an incentive for researchers to gain six months to a year "lead time" over their competitors with respect to new information, it is doubtful that the existing well-documented pedigrees would have been established as quickly as they have been. It is the issue of sharing cell lines after some reasonable time that remains of concern.

Eventually, in the course of the searches for several important genes, competition and cooperation in differing proportions have finally led to discovery. Such has been the case in the successful searches for the Duchenne's muscular dystrophy gene and the cystic fibrosis gene, and the not yet completed search for the Huntington's disease gene.

Ultimately, the sheer difficulty of the gene searches, despite the highly competitive aspect, has often required several or many groups to band together to achieve the final goal. Such I suspect will be the case for at least some of the remaining Alzheimer's disease genes, and the sooner these collaborative efforts can go forward, the better it should be for all of us. Yet, as impatient as I am as a physician to see these genes found, as a basic scientist as well, I do not believe that we can meddle successfully and require cooperation among molecular geneticists unless their mutual interests are also served. We must search for ways to induce time-saving cooperative ventures without subduing the intense efforts and supremely energetic individual initiatives, molded in part by competitive urges, that drive our colleagues in molecular genetics ever forward.

Until then, the interests of my patients and those of my colleagues in molecular genetics remain propelled by different motives. For those at risk for genetic diseases, the principle of mutual aid holds sway and, with it, an understanding that their objective is the survival of their kin if not of themselves. But for the molecular geneticist at the cutting edge of his or her field, there are no immediate concerns for kin selection. His or her own personal survival as a productive scientist is at stake. Indeed, our scientists are often locked in a struggle for the survival of the fittest.

Yet somehow, thanks to the combined strivings of our patients, their families, their physicians, and especially our basic scientists, with enlightened support from funding agencies, public and private, genes are found and a rational basis for the treatment of important genetic diseases has begun. So has it been for Duchenne's muscular dystrophy, cystic fibrosis, and a number of other genetic diseases. So, too, it is beginning now for Alzheimer's disease. Yet with so much at stake, the dilemma as to how best to speed the course of discovery continues to haunt us all.

Hannah's Heirs

Much has happened since that day in May 1985 when a man named Jeff made a chance visit to my clinic. Since then, the search for the mutation that causes familial Alzheimer's disease in Jeff's family has progressed at an ever-increasing tempo. Yet for me, Jeff's visit led to an equally memorable part of this story, a family legacy of enduring courage and determination shared over generations by Hannah's heirs. The inspiration that I have received from the example of her descendants—and especially from Charles and Ben—has been the driving force behind my continued involvement in the genetic research on Alzheimer's disease and my writing this book.

Charles and Ben have survived to witness an ever-broadening and increasingly effective attack on the various causes of Alzheimer's disease. They, as we, have lived to witness the extraordinary scientific and intellectual triumphs of molecular genetics. At no time in history has humankind progressed more rapidly to have the means to find the causes and cures for many dread diseases, perhaps within a decade for some, within a generation for others. Scientific golden ages, each richer than the last, continue to follow one upon another.

Yet there is still no golden age of cooperation among our scientists, in which all available DNA from afflicted patients and their families are shared fully. Accounts of one group refusing to share cell lines with another or providing only limited samples to colleagues continue to circulate in June 1992 as I close this chapter. Misunderstanding between researchers, delays in honoring commitments to share valuable resources, and, too frequently, strained personal relationships between various scientists continue to slow the pace. But such matters, though of vital importance with respect to the conduct of research in human molecular genetics, are not the real story of Hannah's heirs.

The problems of cooperation between scientific groups are not the only difficulties: the very compartmentalization of our work is itself a factor. The neurologist who cares for the patient and his or her family does not usually perform the DNA analyses. The molecular geneticists who work with the clean white crystalline DNA of our suffering patients may labor in research facilities remote from the clinics and nursing homes from whence the DNA samples of our patients were obtained. Milton Wexler had the right idea when at each Huntington's disease workshop he invited all the scientists to meet patients with Huntington's disease and their families, so that the researchers would ever keep in mind the very real suffering of those whom they have dedicated their lives to help. Perhaps following his lead we can go a step further.

Let us as physicians and scientists look into the eyes of our at-risk family members and see them not as numbered aliquots of DNA but as the human beings they are, with their strengths and failings, their hopes and their fears. Let us take the time to follow our affected patients as they fail and join their families in their agony of deciding when to place a loved one in a nursing home. And let us attend a funeral of an afflicted patient and face the question that at-risk descendants often ask, "When will you be able to tell me whether I or my brother or my sister carries or has escaped Dad's gene?" And finally, let us take another moment and peer even more deeply into their eyes and ponder the unstated question that none has yet dared but all must dare to ask, "Doctor, are you and your colleagues doing as much as possible with my family's DNA to bring that day closer when we shall know the cause and cure of our familial disease?"

When we as a scientific and medical community can unflinchingly answer, "Yes," then shall we merit to be counted among Hannah's heirs.

Epilogue

Landfall

Landfall for Hannah's heirs came on September 21, 1992 when Peter St. George-Hyslop decisively linked the abnormal gene among the afflicted of Hannah's family to a marker on the long arm of chromosome 14. Within a six-week period during the fall of 1992, three research groups independently published their discovery of the same site. The DNA samples freely given by the descendants of Hannah and Shlomo and the corresponding neurological and neuropsychological assessments agreed to by the donors of the cell lines made a major contribution to the success of two of these groups—the Seattle group led by Gerard Schellenberg[1] and the Toronto group led by Peter Hyslop.[2] And from Antwerp, Christine Van Broeckhoven's group[3] reported the same linkage for two large Belgium pedigrees with an atypical and perhaps unique variant of familial Alzheimer's disease.[4] The significance of the results reached far beyond their relevance for Hannah's heirs, for within weeks it became evident that the major genetic locus for the vast majority of cases of early-onset familial Alzheimer's disease had now been discovered.

The three independent discoveries of the chromosome 14 site would strike the scientific world with the suddenness of an earthquake, but like earthquakes the conditions creating the groundbreaking developments were long in coming. Gene mappers had eyed chromosome 14 since 1983, when Lowell Weitkamp and his colleagues[5] at NIH had published a paper suggesting a weakly positive lod score at a locus called "Gm" close to the telomere, or bottom end, of the long arm of chromosome 14.

The lod score—the unit for assessing the strength of genetic linkage of a disease to a marker—bears several similarities to the Richter scale used to measure the severity of earthquakes. Both are measured on a logarithmic scale in terms of powers of ten. With each increase in lod score by 1, the odds in favor of linkage go up tenfold. Lod scores of the order of 1 are "interesting," scores of 2 suggest the possibility of linkage, and scores of 3 are taken as the minimal lod score providing statistically acceptable evidence of linkage.[6] A lod score of 7 for a single marker has about the same significance for gene mappers as a powerful earthquake registering 7 on the Richter scale has for seismologists. But geneticists cannot—except by chance—reach for markers that will immediately yield robust lod scores. They must probe their ground carefully in the vicinity of markers that produce the smallest of tremors in the 1.0 range, even while knowing in their hearts that most of these low scores will not lead them to the big one. In retrospect, the Weitkamp result was not an indication that the genetic locus for familial Alzheimer's disease was nearby,[7] but it did arouse a long-standing interest in chromosome 14.

Interest returned to chromosome 14 in February 1988 when Carmella Abraham and her colleagues Dennis Selkoe and Huntington Potter at Harvard Medical School discovered that an interesting enzyme called α_1-antichymotrypsin, abbreviated as AACT, was found in the same plaques that contained the amyloid deposits of Alzheimer's disease.[8] This enzyme inhibited a serine "protease"— another enzyme with an "active" serine site that splits or breaks down certain proteins. The gene for AACT was located on the long arm of chromosome 14.

Abraham's research group suggested several models by which abnormalities in the AACT gene might lead to the development of Alzheimer's disease, and they contacted various gene mappers studying familial Alzheimer's disease, suggesting that these geneticists test the AACT gene for possible linkage. Every group obtained roughly similar results using the relatively uninformative markers then available. Lod scores of only a little greater than 1 were obtained; however, no given allele was consistently coinherited with the disease. This inconsistency of linkage, or "recombination," between a test marker and a putative gene for Alzheimer's disease rendered the odds remote that a defect in the AACT gene itself was a cause of Alzheimer's. Like the gene for the Gm marker, the gene for AACT resides toward the bottom end of the long arm of chromosome 14; unfortunately, there remained a large gap of unchartered territory for regions of chromosome 14 from the AACT gene toward the centromere. Hence, none of the gene mappers could follow the AACT lead any further in early 1988, nor was the evidence compelling that a gene for Alzheimer's resided nearby.

But by late 1991, with no signs of new mutations on chromosome 21 other than those few detected within the gene for the amyloid precursor protein (APP), the various research groups accelerated their efforts to test other chromosomes for the 97% of families with Alzheimer's disease without mutations in the APP gene. The preferred technology for searching for mutations had by now changed rapidly. The RFLP markers upon which the gene mappers of the 1980s had relied had given way to the short tandem repeat (STR) markers, which were highly polymorphic and thus extremely informative. In 1989, James Weber and Paula May proposed that this new class of markers would open up vast new opportunities for genetic linkage studies.[9] The Human Genome Project was now heavily funding a number of laboratories to generate as rapidly as possible the highly informative linkage maps of various chromosomes based on these STR markers.

It was scarcely three years since Weber and May had first suggested that certain unique blocks of DNA recurring in repetitious sequences scattered throughout the genome could serve as highly informative genetic markers. Yet their discovery had already revolutionized their field. Sequences of nucleotides such as cytosine, adenine, cytosine, adenine, . . . , abbreviated as CACA . . . , could recur numerous times, and the gene mapper had only to measure the length of the repeat to characterize one "allele," or one form, of the marker at a given chromosomal site. These new STR markers were extraordinarily informative. Unlike the RFLPs, for which there often existed only two different lengths of markers at a given site, the variations in lengths of the STRs were much more numerous, with 8 or 10 or even more variable lengths often present at each marker locus. With so many alleles present at one location, it became much easier for gene mappers to trace the association between an allele and a genetic disease throughout a family. For example, if a marker found in an affected individual occurred with only a 10% incidence within the general population, it was unlikely that both parents would carry this allele. Hence, gene mappers could determine with high accuracy whether a given allele was inherited exclusively from the affected parent.

Geneticists could readily measure the lengths of the STRs after they discovered the unique sequences of short strips of nucleotides "bracketing" the STRs at each end. Geneticists could synthesize these sequences called "primers," then use the polymerase chain reaction to amplify only that sequence of DNA interposed between the primers, and finally measure the length of the amplified marker DNA. The length of each strip is directly proportional to the number of (CA) blocks present and can thus be used to define each allele.

For the first time, geneticists had more chromosomal markers available to them than any one group could test independently in a reasonable time. And since the research groups were for the most part neither sharing pedigrees nor working collaboratively, most gene mappers fell upon a strategy of testing at least one highly informative marker for each arm of every chromosome while working in relative isolation from each other. Present-day research in the genetics of Alzheimer's disease has become about as competitive as research can get.[10]

The Seattle group was the first to publish their evidence for linkage of familial Alzheimer's disease to chromosome 14. They tested over 60 markers, scanning across the genome in intervals that were as optimal as possible based on the availability of informative markers, but the decisive factor for their success came when Gerry Schellenberg, who scoured the world's literature for useful new markers monthly, happened on a crucial one-half page account of the discovery of a new marker in a paper by Vikram Sharma and his associates at the Oregon Health Science University in Portland. The account appeared with the seemingly bland but succinct title "Dinucleotide repeat polymorphism at the D14S43 locus" in the journal *Nucleic Acid Research*, which was published in April 1991.[11] In a terse single-sentence summary, the authors suggested that "this marker locus will be a useful index marker for the genetic linkage map of chromosome 14." Schellenberg selected this marker as his screening marker for chromosome 14. Upon finding decisive linkage to this marker, he contacted James Weber who was known to be constructing a linkage map for chromosome 14. Schellenberg learned of the existence of Weber's new markers for chromosome 14, tested these, and mapped the D14S43 marker in relationship to these other markers.

For James Weber, the decision to construct a linkage map for chromosome 14 had been first motivated by two factors. He told me that chromosome 14 had been a largely neglected chromosome because of the lack of useful markers for an extended section of the long arm. Moreover, by chance, he had already found a number of markers that mapped to chromosome 14. Hence, he felt that the time was right to systematically apply the methods he had developed to make a map of short tandem repeat (STR) markers for the entire long arm. Fortuitously, his map was largely complete by the time Schellenberg contacted him.

The discovery of the key D14S43 marker is itself a remarkable example of scientific serendipity. Schellenberg alerted me to the story, which Michael Litt of the Department of Medical Genetics at the Oregon Health Sciences University in Portland then elaborated. In the 1989–1990 period, there was great interest in finding informative markers for chromosome 21 in regions flanking the first markers which the Hyslop group had then reported to link to at least one form of familial Alzheimer's disease. Litt had a young graduate student named Vikram Sharma who wanted to find informative STR markers to search for an Alzheimer's disease gene on chromosome 21. He obtained what is called a "library," containing fragments of DNA from chromosome 21, and extracted a probe that to his disappointment "hybridized" or attached only to chromosome 14. In other words, the probe was a contaminant of the chromosome 21 library. Litt told the disheartened Sharma, "You have found something more important. You have filled the 'black hole' in the long arm of chromosome 14." Of course, at the time neither Sharma nor Litt had any inkling that the marker would be the key to the major locus for familial Alzheimer's disease.

The paper by the Seattle group was published in the October 23 issue of *Science*. Schellenberg's group convincingly linked a pedigree of German ancestry to the D14S43 marker with a peak lod score of 4.89. Additional suggestive evidence for linkage to the same region came largely from two other families. In the mid-1980s, Schellenberg had given Hyslop samples of DNA from one affected member and one elderly escapee in Hannah's family and Hyslop had given Schellenberg samples of the DNA from the remaining other members of Hannah's family, whose status was then known. Using these samples, Schellenberg obtained a lod score of 2.17 for linkage of Hannah's family to the D14S43 marker. Schellenberg's other noticeably positive lod score of 2.32 came from a French Canadian pedigree, identified as the LH (603) family, which Allen Roses had extensively studied. Schellenberg had independently studied a small branch of the family living on the West Coast and received cell lines from members of the East Coast branches from Allen Roses. However, Schellenberg's lod scores for these two families did not reach the minimum value of 3.0 required for establishing linkage. (Ironically, the Volga German families, long studied by the Seattle group, showed decisive evidence against linkage to chromosome 14. Hence, there must exist at least one more chromosomal locus for early-onset familial Alzheimer's disease apart from those for the defective genes on chromosomes 21 and 14.)

Schellenberg's group had established evidence for linkage of familial Alzheimer's disease in one family to chromosome 14 and provided suggestive evidence that two others might link to this site as well. It remained to be determined whether the chromosome 14 linkage site would be one of numerous rare chromosomal locations for Alzheimer's disease genes or might just be the major locus. This answer was already known to Hyslop's group, who had independently arrived at the same observation and discovered that the linkage held for a sizable number of families.

For Hyslop's team, the path to the key marker on chromosome 14 had been long and tortuous. Just as was the case for the Seattle group, Hyslop's Toronto group tested over 60 probes before they found evidence for linkage to chromosome 14, but Hyslop remained unaware until relatively late of the existence of the D14S43 marker. In May 1992, Marzia Mortilla, one of his postdoctoral fellows, telefaxed a note to Weber requesting information about new markers for chromosomes 13, 14, 16, and 18. While awaiting the reply, the Toronto group began testing two markers toward the centromere of chromosome 14 called D14S50 and D14S52. The disease failed to link to these markers.

Then, in mid-July 1992, the Hyslop group obtained Wang and Weber's linkage map of the entire long arm of chromosome 14 based on STR markers which had just been published in the journal *Genomics*.[12] Now, with the Wang and Weber primer sequences, Hyslop's group immediately began synthesizing the remaining five pairs of primers needed to cover the entire linkage map of chromosome 14. By month's end, Hyslop gleaned his first hint of noticeably positive lod scores using the D14S48 marker in FAD1, the British Canadian family, testing the affected-only data set. He had not yet proved "linkage," the lod score was only close to 2, but the score encouraged Hyslop that he might be getting close. Hyslop now had to test DNA samples from over 400 people for each test marker, and he had to formulate a strategy to reduce unnecessary work. Among his pedigrees there were 97 samples from affected individuals, the remainder were from "escapees" and individuals still at risk. He reasoned it would be initially faster to test the affected-only population, and if a marker were negative, then he would go on to another. His plan was that only when he obtained very significant positive results from studying the affected-only members of a pedigree would he type the escapees as well. The result on the British Canadian family was tantalizing, but not yet strong enough for him to start testing the escapees.

On September 10, 1992, Hyslop's group discovered their first statistically significant evidence of linkage to chromosome 14 in a pedigree of Italian descent identified as TOR1.1 (TOR for Toronto) using the marker D14S53. The lod score was convincingly positive at 5.38, but the discovery created a perplexing dilemma. The Calabrian pedigree from southern Italy designated FAD4 was thought by some workers to share a common founder with the TOR1.1 pedigree, and yet the lod score for the very large FAD4 pedigree was barely greater than 2. And until this time, most gene mappers suspected that the Calabrian pedigree would eventually link to chromosome 21 because previous lod scores for a chromosome 21 marker not far from the amyloid gene had consistently registered in the mid-2 range—although never above the 3.0 taken as the conventionally accepted value for asserting that linkage has been demonstrated. Hyslop's group then tested another marker, D14S55, which is on the telomeric side of D14S53, and discovered robust linkage for the Calabrian pedigree. The evidence for linkage of this family to chromosome 14 was now more than one hundred times greater than the previous suggestive but never definitive scores in favor of linkage to chromosome 21. Now the two pedigrees of Italian origin—or one kindred if both families were descended from a common founder—were linked to chromosome 14.

But what of the other families? Would these also link to chromosome 14? Some of the atypical characteristics of Alzheimer's disease in the Calabrian family—the very early onset of the disease, frequently when the victim was in his early thirties, and the often accompanying seizures—continued to raise the question of whether their disease was caused by a mutation in the same gene that caused the more classically described forms of the disease seen in most other pedigrees.

Hyslop scrutinized Weber's paper again and found a reference to a probe identified as 2E12b, but without a D-number used to assign a specific chromosomal site. This new probe was very closely linked to the D14S53 marker, but the relative position of the marker—that is, whether it was centromeric or telomeric to D14S53—was not reported. Hyslop's group consulted the Genome DataBase and finally found the detailed description of new probe now designated as D14S43.

The primers were made on September 14, and within a week, Hyslop's group decisively linked Hannah's family (lod score 6.99) and the British Canadian family (lod score 3.99) to the D14S43 marker. All affected individuals in the two families linked to the marker. According to the Wang and Weber paper, the D14S43 and the D14S53 markers were linked so closely that they became separated only 2% of the time during the chromosome shuffling or chromosomal cross-over that occurs during meiosis. Thus, in the parlance of the geneticist, the markers were separated by 2 "centimorgans"—the term coined in honor of Thomas Hunt Morgan used to define the frequency of such cross-overs.

After over seven years of work, Hyslop had now within a few weeks linked four large pedigrees to the same general region of chromosome 14. It still could not be determined whether the Calabrian pedigree and Hannah's family linked to adjacent genes or involved different mutations of the same gene, but the generality of the chromosome 14 site as a major locus for familial Alzheimer's disease was rapidly becoming apparent. Hyslop wrote up the results for submission to *Nature,* mailing off his draft to the London office of *Nature* on September 30.

Before completing the final draft, Hyslop phoned Allen Roses at Duke, told him of the independent linkage of all four pedigrees to the chromosome 14 markers, and invited him to join the authorship inasmuch as the two had agreed previously to work together on their screening of the genome. Roses congratulated Hyslop but declined joining since he had not as yet contributed to the new discovery. Roses agreed, however, that he would test his large pedigree identified as "603" with the probes that Hyslop had prepared; if his results were informative, he would join the authorship when the paper came back from the reviewers. (Roses's 603 family is the same "LH 603" pedigree that would soon be included in Schellenberg's report.) So powerful and so speedy are the new techniques utilizing the polymerase chain reaction (PCR) for testing linkage to the new short tandem repeat markers that within only several days Roses and his colleague Margaret Pericak-Vance obtained linkage (lod score 3.13) for pedigree 603 to one of the chromosome 14 markers, and they joined Hyslop's paper.

Up to this point, Hyslop had tested only Hannah's family and the British Canadian family to the D14S43 marker. In fact, even up to the time of submission of his paper to *Nature*, Hyslop had not realized that the D14S43 marker was the key marker. While waiting to hear from *Nature*, the Toronto group decided to test the remaining pedigrees with this marker. The German pedigree (FAD2) now linked with a lod score of 3.65.

Within less than a month, the Toronto group—with an assist from Roses and Pericak-Vance—had linked six of the world's largest and best documented pedigrees to closely adjacent chromosome 14 markers with individual lod scores of greater than 3. It was as if Hyslop's team had detected not one earthquake but six. The lod score for Hannah's family was the most compelling of the six, with a value of 6.99—a virtual 7. In fact, the cumulative lod score for just the D14S43 marker in the six pedigrees that linked to chromosome 14 was 19.16, and the marker and the disease were inherited together in every case. Thus, the abnormal gene, or complex of several abnormal genes, must lie very close to this marker. (See linkage map on page 250.) An expert interviewed later by Celia Hooper for her article on the linkage discoveries for the *Journal of NIH Research* cited the Hyslop team's paper as "the most stunning piece of linkage data in the field."[13]

The twinfold significance of the linkage discoveries of these three research groups were plainly evident. That nine large independent pedigrees linked to chromosome 14 indicated that this chromosome must comprise the major locus for the majority of families with early-onset familial Alzheimer's disease. Second, and more important, the new gene locus was entirely distinct from the amyloid gene found on chromosome 21. Thus, as soon as the new gene(s) could be cloned, modern genetics would be poised to answer a crucial question—were the genetic defects on chromosome 14 causing Alzheimer's disease by affecting some aspect of amyloid metabolism, or were they producing the disease by mechanisms entirely distinct from those involved in the metabolism of amyloid? From this knowledge, the decisive information about the ultimate causes of Alzheimer's disease would be uncovered, further

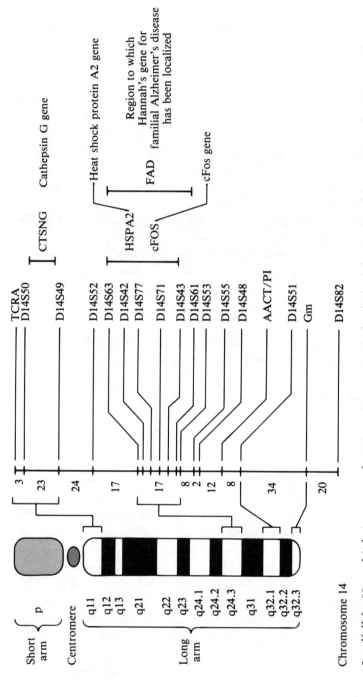

Chromosome 14

Landfall for Hannah's heirs came when Peter St. George-Hyslop linked the genetic defect among the afflicted of Hannah's descendants to the D14S43 marker shown on this genetic linkage map of chromosome 14. The genetic defect in Hannah's family is most probably located within the q24.3 band (bands indicated to the left of the schematic long arm of the chromosome) and most likely between the markers D14S53 and D14S63 (markers denoted to right of chromosomal representation). (Courtesy of Peter St. George-Hyslop.)

opening the window to our understanding of the disease process in its various forms.

Since mid-July 1992, when Hyslop's group received the Wang and Weber paper, the group had thoroughly studied 21 pedigrees with eight markers and run over 3,000 sample runs. (Many of these smaller pedigrees may also eventually link to chromosome 14. In many cases, the absence of strong evidence for linkage in a small family simply reflects the fact that the pedigree structure is not large enough or informative enough for linkage to be demonstrated.) Worn out by the effort but exhilarated by the results, and confident of a rapid acceptance by *Nature*, Hyslop's group now turned to see what known genes lay close to the newly defined locus. Ironically, the AACT gene, which had initially sparked the attention of all groups to chromosome 14, appeared too far away to be a serious candidate gene.

Although the AACT gene seemed out of the running, the D14S43 marker was very close to the C-Fos gene—a gene whose protein product can modify the expression of numeous other genes—and also close to a reported site for a gene that expresses a "heat shock protein" called HSP 70. The heat shock proteins are so designated because their synthesis increases under heat stress. Under more normal conditions they may be involved in the assembly of other proteins, guiding or "chaperoning" certain proteins to essential locations within cells. Hypothetically, abnormalities in such chaperone proteins might lead to the death of nerve cells, as in Alzheimer's. Thus, the HSP gene immediately became an exciting candidate gene—all the more so in view of the March 1991 report by James Hamos, Barry Oblas, and David Drachman of my department at the University of Massachusetts Medical Center that these proteins increase in Alzheimer's disease and seem to be found exclusively in the neuritic plaques and neurofibrillary tangles.[14]

Yet Hyslop's expectation that his linkage results would be published in the widely read journal *Nature* was not fulfilled. A reviewer recommended on narrow technical grounds that the new manuscript not be published in *Nature* but rather be resubmitted to *Nature's* new specialty journal *Nature Genetics*. Thus, Hyslop's paper was subsequently published in the December 1992 issue of *Nature Genetics*.

By early January 1993, Hyslop's team had ruled out mutations in the coding regions, or exons, of the C-Fos gene as causes of Alzheimer's disease in several of the large pedigrees. The full

sequence for the heat shock protein gene and even its precise location on chromosome 14 have not yet been determined as of early
1993, but its screening continues, and may take appreciable time.

As 1993 began, there were nine large pedigrees that showed
independent, statistically significant linkage to Sharma's D14S43
marker on chromosome 14—six from Hyslop, two from Van
Broeckhoven, and one from Schellenberg. The cumulative lod score
for linkage for these nine families was now over 40, and in every
case the genetic defect for Alzheimer's disease and a telltale allele
of the D14S43 marker have been inherited together. In fact, Van
Broeckhoven's group suggested that the D14S43 marker may be
contained within an Alzheimer's disease gene in view of both the
profound evidence of linkage to this marker, with no yet observed
discrepancies or "recombinations," and the fact that the short tandem repeats are often found within the "introns," or the noncoding
intervening sequences, of expressed genes. Of course, it is equally
possible that the D14S43 marker is part of a gene very close to the
Alzheimer's disease gene(s). Soon, perhaps within a year or two,
thanks to the pioneering efforts of these three research groups, the
gene or genes on chromosome 14 responsible for the majority of
cases of early-onset Alzheimer's are likely to be identified.

That day when the precise genes are known cannot come soon
enough. In early October, I phoned Charles to convey some of the
news of Hyslop's discovery. He immediately grasped its
significance at several levels. As a loving grand-uncle, Charles realized immediately that the fate of his at-risk grand-nephews and
grand-nieces could now be predicted with very high accuracy by
genetic testing before symptoms appear. Charles was as close to
some of these young adults as any grandparent could be, but he
knew that the information as to their status was privileged and
could be made available only to those family members who
requested testing. It would become my responsibility to notify each
of the at-risk members of the fifth generation that such a test had
now become available. Charles asked Bryna to update the addresses
and phone numbers of all at-risk, fifth-generation and potentially
at-risk, sixth-generation family members, so that in time all necessary information about the discovery and its implications for their
lives, and that of their families, could reach them.

Of course, neither Charles nor I would take it upon ourselves
to recommend that at-risk family members request genetic testing.
That decision would be a personal and family matter for those

involved. The issues surrounding the predictive testing of at-risk persons for diseases that have no present cure are so complex and so difficult to resolve from an ethical point of view that I cannot address these matters here. Even so, a study recently conducted of individuals confronted with the choice of accepting or not accepting presymptomatic testing for Huntington's disease[15] found, as expected, that those who tested negative for the gene experienced a profound sense of relief and led more productive lives than they perhaps would have otherwise. But equally important, the individuals who tested positive for the genetic defect of Huntington's carried on surprisingly well emotionally in the years prior to the onset of symptoms compared to those who chose not to be tested and to live their lives in continued uncertainty.

However, with respect to Hannah's heirs, I actually hoped that no one would request testing, at least for now. First, until the actual Alzheimer's disease gene is found and its DNA sequenced, the accuracy of any test cannot be 100%. By early January 1993, the accuracy of a predictive test was probably over 95%, and it will increase rapidly as gene mappers close in on the defective gene. Second, I strongly believe that, once the gene is known, a fuller understanding of the mechanism by which it produces Alzheimer's disease will follow within a few years, and rational treatment of the disease will commence. I believe that if we have reached the present high ground in understanding familial Alzheimer's disease, within scarcely more than 40 years since the Watson and Crick discovery of the double helix, that it is almost inconceivable that we will not have a viable, if not perfect, treatment for the consequences of these genetic defects within the next 40 years. But my personal beliefs in the likely progress of this science must not dictate the choices of the at-risk descendants of my patients.

For those at risk who wish to become new parents in the next several years, the choice of whether to undergo testing to determine their own status, and perhaps even subsequent prenatal testing of their prospective offspring, presents a decided dilemma. For those who may test positive for the marker, the accuracy of a test for determining whether or not a fetus carries Hannah's gene for Alzheimer's probably just exceeds 90%.

It will not be an easy choice for anyone, but it is a choice, and one that Ben and Charles did not have when they decided to forgo bringing children of their own into the world. As for some of Hannah's other descendants, Hyslop's discovery has already changed their lives. Some fifth-generation, at-risk members who had decided that they would never marry—or if they did marry would not have children—have now chosen to marry and will have children if they can be assured with reasonable certainty that the fetus does not carry the Alzheimer's disease gene.

Apart from his roles as devoted advocate and trusted advisor for his extended family, Charles, as scientist, has once again begun to think of new approaches that might help zero in on the gene even more rapidly. Charles was familiar with the work of Dr. Jeanne Lawrence, an investigator at our medical school, who had become well known as a ranking expert on "in situ hybridization" methods. These methods use fluorescent tags that are attached to DNA probes that are then visualized by light microscopy after they bind selectively to a chromosomal site.[16] Charles thought we might invite Dr. Lawrence to carry out such studies on chromosome 14 using an affected member of his family and probes that map closest to the genetic defect, to determine whether there is an obvious deletion or duplication of DNA in a region near the probes. Such a finding would signal that one or the other of two possible types of genetic defects might be involved.[17] Dr. Lawrence agreed to carry out the studies, and this work is now under way.

Meanwhile, several new probes close to the region of greatest interest on chromosome 14 have recently been developed and have been added to the map of Wang and Weber. As of early February 1993, none of the possible candidate genes has yet been identified as a gene responsible for familial Alzheimer's disease. However, with the genetic defect(s) already so well localized in so many large pedigrees, with such great interest worldwide in the problem, and in consideration of the even now rapidly improving technologies for sequencing long strips of DNA, many workers believe that the chances are high that we will see the chromosome 14 gene(s) discovered within several years at most.

And as for the more prevalent late-onset Alzheimer's disease, there are some promising recent genetic results as well. Allen Roses's group now has strong but still not yet conclusive evidence for a gene on chromosome 19 that might confer susceptibility to the late-onset form of the disease. One of the genes in this region codes for a lipoprotein called apolipoprotein E. Roses's group has also found that 50% of patients with an apparently familial form of late Alzheimer's disease have also inherited the type 4 allele of this lipoprotein as compared with a frequency of about 15% in age-matched unrelated controls. An increased frequency of this allele is also found in late-onset sporadic Alzheimer's disease, which occurs in the population at large. Going a step further, Roses's group suspects that the increased risk of Alzheimer's is related to strong chemical binding between β-amyloid and the type 4 lipoprotein.[18] If all of these observations are confirmed, then there are a number of implications for eventual treatment of patients with late-onset Alzheimer's disease, the form believed to affect from 3 to 4 million people in the United States alone.

As I look back over the past seven and a half years since Jeff's visit to my clinic, I sometimes think of the many leads followed, the ones that worked, the ones that did not, and some of those which did not work that led us to others that did. The beacon that had led all gene mappers to the long arm of chromosome 14 had been Carmella Abraham's discovery of the enzyme AACT in the cerebral plaques of patients with Alzheimer's disease, and the AACT gene's location toward the tip of the long arm of chromosome 14. Yet in the end, it turned out that the major locus for the familial Alzheimer's disease genes, though on the long arm of chromosome 14, was not likely to include the AACT gene.

In retrospect, my search for Hannah's descendants in Russia failed to reveal any new affected cases, although it did lead us to discover a large branch of the family who happily escaped the abnormal gene and who informed us as to the family's prior history in Lithuania. Unfortunately, the search by our colleagues in Lithuania has so far failed to discover any common descendants of the ancestors of Hannah and Shlomo who shared an early-onset Alzheimer's disease gene. And in retrospect, the significance of my trip to Russia in 1988 was not the samples from the Russian pedigrees that I brought back to Peter Hyslop, but the opening I established for further contact between the research teams of Ivan Alexandrov and Yuri Yurov at the then All Union Mental Health Research Center and Hyslop's group. Following upon my trip to Moscow, Hyslop visited the research group in 1989 and met Evgeny Rogaev, a member of Yurov's group. Rogaev expressed an interest in working with Hyslop, then still at Massachusetts General in Boston. Thanks to a research fund that Jeff's widow Susan had established after Jeff's death, funding became available for Rogaev to come to Boston for a brief working visit. In February 1992, Rogaev arranged to return to Hyslop's lab, now in Toronto, for an extended visit. Rogaev became a mainstay of the Hyslop laboratory in the discovery of linkage to chromosome 14 and was first author on a subsequent paper eliminating coding regions of the C-Fos gene as candidates for the Alzheimer's gene.[19] Susan's commitment to research, just as Jeff's, had already quickened the pace of discovery.

But of all the paths taken, none has remained as well marked as those forged by the dedicated efforts of Hannah's descendants. Soon, very soon I hope, their genetic defect will be found. With that day I, as well as all of my colleagues in neurology, will look toward further advances in understanding a major cause of Alzheimer's disease and be ever closer to treatment. But the first part of this story—the quest for the genetic origins of Alzheimer's disease—is now well along with the discovery of linkage, and I can think of no better words to capture the spirit of this struggle than those of Ben: "This is a story that had to be told. Our aspirations were transcendent, but because it involved people, it could not be told without tears."

It is still a dark night for all those who suffer from Alzheimer's disease, and much work lies ahead, but surely we are now long past the dark of midnight and reach with confidence toward that brightening which presages dawn and promise.

February 5, 1993

The finding by Allen Roses's group at Duke that the variant of apolipoprotein E called apoE-E4 is a major risk factor for late-onset Alzheimer's disease was swiftly and repeatedly confirmed.[20,21] The risk of developing Alzheimer's increases by roughly threefold for the 31% of the population inheriting at least one copy of the apoE-E4 variant and by eightfold for the 2% of the population inheriting two copies compared to the risk for those inheriting no copies.[22] The increased risk expresses itself most frequently during the sixties and seventies and essentially decreases the mean age of onset of Alzheimer's disease for those who inherit one or two copies of apoE-E4 by 9 and 17 years, respectively, compared to the mean age of onset of 85 for those who inherit no copies of apoE-E4. Women who inherit one copy of apoE-E4 may bear a disproportionate burden of the risk than do men, although the eightfold increase in risk for those inheriting two copies of the E4 allele holds equally well for men and women.[21,23] Conversely, one of the three forms of apoE, the apoE-E2 variant, appears protective. The approximately 8% of the population inheriting one copy of this variant have a substantially reduced risk of developing late-onset Alzheimer's and even those affected are more apt to develop it at advanced ages.[24,25]

The mechanisms by which the apoE-E4 variant confers

increased risk of Alzheimer's remain intensely controversial. Even before genetic studies identified the apoE-E4 variant as a specific risk factor, Yoshio Namba's team[26] in Tokyo had unexpectedly discovered that antibodies to apoE bound strongly to the β-amyloid in senile plaques and neurofibrillary tangles. Shortly thereafter, Thomas Wisniewski and Blas Frangione[27] proposed that the binding of apoE to β-amyloid facilitates the conversion of β-amyloidís diffuse, soluble, and presumably harmless form to a fibrillar, insoluble, and presumably neurotoxic form. Then rapidly following upon their genetic discovery, the Duke workers found that it was the apoE-E4 variant that bound most strongly to β-amyloid.[28]

Meanwhile, evidence had accumulated that β-amyloid fragments are derived from the β-amyloid precursor protein in several lengths and that the longer 42 or 43 amino-acid long residues comprise the earliest deposits and most characteristic amyloid component in senile plaques.[29] The longer fragments tend to aggregate spontaneously into amyloid fibrils,[30] a process that Huntington Polter's group at Harvard believes is selectively accelerated by the presence of apoE-E4 and also by α_1-antichymotrypsin (AACT).[31] Many workers believe that these insoluble fibrils then trigger a series of both direct and indirect[32,33] toxic effects on nerve cells. Thus, proponents of amyloid models of neurotoxicity, as exemplified by Dennis Selkoe, anticipate that the development of drugs to prevent the aggregation of β-amyloid into fibrils and then plaques offers the best chance to prevent or slow the development of Alzheimer's.

Even so, Roses suspects that the binding of apoE to the tangles found inside nerve cells in Alzheimer's may be more fundamental to the disease process than its binding to the β-amyloid found outside such cells in the plaques.[34] His team has suggested that variants of apoE differentially bind to microtubule-associated proteins such as "tau" inside nerve cells, which then either protect or fail to protect the function and stability of the microtubules. In their view, a compromise in microtubule function in those inheriting an allele for apoE-E4 leads to cell death and the concurrent formation of neurofibrillary tangles that characterize Alzheimer's disease even more specifically[35,36] than do the β-amyloid plaques.

The two main hypotheses proposed for the mechanism of action of apoE-E4 are not mutually exclusive. Other possibilities remain open as well. ApoE can itself form amyloid-like fibrils which hypothetically could seed β-amyloid deposition.[37] Resolution of these conflicting views should lead to an understanding of the disease process which, in turn, will likely lead to the development of rational therapies to delay the onset and slow the progression of late-onset Alzheimer's disease and perhaps in time to avert it. The search for as yet unknown genetic and environmental risk factors offers further promise for the understanding and treatment of late-onset Alzheimer's. Moreover, there may exist still to be discovered age-dependent changes in brain function that increase susceptibility to known risk factors.[38]

February 15, 1996

Hannah's Gene

The discovery of the major gene for early-onset Alzheimer's disease burst forth with the seeming suddenness of a lightning bolt—coming about two-and-one-half years after the gene's linkage to chromosome 14. The race to its discovery had run on past a decade, arduous throughout, fiercely competitive, and unnervingly tense when episodic false rumors flashed through the research community that one or another group had already found the gene. At times, even the integrity of the race was threatened by ruthless assaults on accepted standards governing the conduct of competition in scientific research. Even so, in the end, the discovery was recognized as of such profound significance for the future of research in Alzheimer's disease that assuredly the quest had not been in vain.

By late 1992, using genetic linkage results from the largest families afflicted with Alzheimer's, scientists had narrowed the location of the unknown gene to a region at least five million nucleotides long. For Peter Hyslop's group, one boundary had been established by a recombination event in an affected member of the British-Canadian family. Another genetic cross-over event in an affected descendant of Hannah set the other limit. Various research groups soon eliminated the several genes already identified within this interval as possible sites for the Alzheimer's mutations. The arduous search for an unknown gene had now to be undertaken.

Testing a known gene for a suspected mutation, as in the case for the amyloid gene on chromosome 21, is a straightforward and relatively rapid task. But isolating, identifying, and testing novel genes within an interval as long as five million base-pairs is an entirely different matter. Researchers either had to pull out genes at random or narrow the search to an even smaller region. Fortunately, the discoverers of the genes for cystic fibrosis[1] and Huntington's disease[2] had successfully applied a clever stratagem to narrow their searches by scrutinizing extended "haplotypes" on the affected chromosome across different families of the same ethnic origin. The term "haplotype" refers to the specific combination of very closely linked alleles (alternate subtypes of an expressed gene) or other DNA markers that tend to be transmitted as a unit from one generation to the next. Geneticists can analyze haplotypes to trace the descent of individuals or branches of families back to a common "founder." To accomplish this, scientists scan their linkage maps to see if the affected members of different families share short sequences of identical alleles or DNA markers. Inherited correspondences in these sequences would suggest a common founder affected with the mutation and selective linkage of the genetic defect onto the shorter strip of DNA, which is coinherited in two or more branches of a presumed superfamily. Thus, the genetic record can outdistance the historical limits of geneology, leaving indelible, ancient clues to common ancestors bearing specific mutations.

When Hyslop's team compared the extended haplotypes for the affected members of Hannah's and Elizabeth's families, the researchers recognized that the stretch of chromosome encasing the defective gene was identical with respect to the observed alleles for eleven consecutive markers that spanned an interval of almost the entire five million base-pairs. Thus, Hannah's and Elizabeth's great-grandmothers must have been closely related; otherwise, such long stretches of DNA should not have survived the shuffling of DNA between paired chromosomes during formation of the germ cells over so many generations. Although the comparison of haplotypes confirmed our long-held suspicion that these two families were descended from a single founder based on their common geographic and ethnic background, the combined results could not further narrow the search for the critical region. Nor did the analysis of extended haplotypes reveal any genetic relationship between the founder in Hannah's family and any other family.

Fortunately, Hyslop's team discovered two distinct but closely spaced clusters of identical markers that were coinherited with Alzheimer's disease in the large Calabrian family he had studied and two smaller families also of Italian origin. If the assumption of a common remote ancestor held true, then the gene had to be within or just between the two physical regions defined by the clusters. The affected members of the large British-Canadian pedigree and a smaller family of English origin also shared a sequence of several identical markers, again pointing to one of the regions implicated by the Italian families. The gene had now been narrowed to a region of almost two million base-pairs, still very large but more tractable.

The geneticists could squeeze information provided by family studies no further. Researchers would have to apply other techniques to recover "transcripts" from probable genes out of the newly defined region and examine every novel gene for possible differences in nucleotide sequence taken from normal and diseased chromosomes.

This is where Hyslop stood during the summer and early fall of 1994. Funding for his research was running precariously low and Hyslop, now planning for future collaborative work with Allen Roses to speed the search, had to divert months from his own experimental effort to prepare a detailed application for a research grant for submission to the National Institutes of Health by October 1. Hyslop would use a positional cloning method to isolate and screen candidate genes from the critical band of chromosome 14, identified as 14q24.3, while Roses would provide candidate genes that mapped to 14q, which were discovered while working out the biology of apoE in Alzheimer's disease. While waiting to hear from NIH, Hyslop, Robin Sherrington, a young geneticist from London, England, Evgeny Rogaev, still on leave from his institute in Moscow, and Paul Fraser continued to spearhead the Toronto-based effort to clone new genes based on a technique developed by one of their colleagues, Johanna Rommens.[3]

Her technique first exploited the fact that small fragments of messenger RNA could be recovered from previously frozen brain tissue by being warmed to a suitable temperature. These fragments could then be collected and mixed with the enzyme "reverse transcriptase" that uses the mRNA as a template to generate copies of complementary DNA. The cDNA fragments could then be hybridized to pieces of genomic DNA from the region of chromosome containing the suspect gene, allowing candidate genes encoded on chromosome 14q24.3 to be separated from all other

30,000 genes expressed in the brain but which are encoded on other chromosomal regions.

The effort required was prodigious. As a first step, the Toronto team recovered about 900 cDNA fragments referenced to the suspected location of the gene. Of these, a still large subset of 151 clones had the appropriate characteristics of genes rather than of noncoding DNA sequences. These fragmented segments were then used to screen the cDNA "libraries" to recover larger sequences. At last, nineteen independent gene clones were isolated and mapped to the critical regions. Of these, only two corresponded to known genes. Of the remaining clones, seven transcripts were sequenced either because preliminary screening techniques detected differences between transcripts from normal and affected subjects or because a plausible biological role for a new gene in Alzheimer's disease could be postulated from the inferred amino acid sequence of the protein encoded in the candidate gene.

By Christmas 1994, the Toronto group found a change in a single nucleotide in a DNA transcript from a small frozen specimen of brain tissue that Jeff had willed to be donated after his death in 1989. This change was not observed in brain tissue from any of three neurologically normal subjects initially examined. This event was not the first time that such differences had been observed. In initial studies of candidate genes such changes may turn out not to be unique for all affected subjects and thus do not represent the targeted mutation. At other times, changes had been found in "untranslated" regions of genes and consequently were not expected to change the gene's protein. Tentative hopes of a candidate gene had often vanished as additional transcripts from normal and affected subjects were tested. This time the result held. By early March 1995, the single base mutation in a hitherto unknown gene—designated as clone S182—could finally be identified as the cause of Alzheimer's disease in Hannah's family.

As work in Toronto proceeded to test this same gene for mutations in other families, Roses and Hyslop were informed by NIH personnel that their grant application had been "triaged" or relegated to the bottom half of submitted grants based on the averaged recommendation of only two reviewers, and thus automatically disqualified for funding. One anonymous reviewer, who by the text of his or her critique had intimate knowledge of the field, had summarily dismissed this application precluding further review by the entire study section for reasons that appeared to Roses and Hyslop without substance. Despite the inappropriate dismissal of their funding—the lifeblood of any research lab—Hyslop's team pressed ahead. For the moment, funding was adequate but only for the next several months. Had the loss of their research funding come before the discovery of the gene, the outcome might well have been different and the recent advances propelling research in Alzheimer's disease may have been much delayed.

With the cloning of the gene completed, the team had identified five distinct single nucleotide or "missense" mutations—one error or single nucleotide change in each of their five largest families of different ethnic origins—at sites widely spaced along the new gene. The team could now deduce the structure of the expressed protein from the normal gene's nucleotide sequence. The deduced gene product consists of a 467-amino-acid protein predicted to contain at least seven membrane-spanning "domains." Membrane-spanning proteins traverse a membrane and extend beyond the limits of both the interior and exterior membrane surface. The S182 protein looped back and forth through the membrane at least seven times.

But how might this new protein, which they subsequently called "presenilin 1," function in normal subjects and, when abnormal, cause Alzheimer's? Based on just its structure, the S182 protein might be an integral membrane protein, a "receptor," or a "channel" for ion flow. Perhaps there would be a further clue as to the function of S182 if its nucleotide and amino acid sequence could be compared with those sequences of all identified proteins stored in a central database. Surprisingly, S182 bore a close resemblance to a membrane-spanning intracellular protein found in the sperm of a flat worm known as *C. elegans*. The sperm protein, identified as SPE-4, is probably involved in the storage and transport of soluble and membrane-bound polypeptides. Moreover, Hyslop's team soon found that the S182 protein is also located intracellularly. Thus, comparable functions could be envisaged for the S182 protein. Perhaps the mutated protein interacts abnormally with other membrane-bound proteins such as the β-amyloid precursor protein or with the microtubule-associated protein "tau."

With the discovery of Hannah's gene, the bottleneck preventing further understanding of the pathogenesis of the majority of cases of early-onset Alzheimer's was now broken. The way was now open for scientists to discover how the new protein interacted with the amyloid precursor protein, its possibly neurotoxic β-amyloid fragment, the microtubule-associated proteins, apoE and still other proteins, to unravel the underlying sequence of events leading from a change in a single nucleotide to a devastating disease.

At last, on May 15, 1995, Hyslop sent off the manuscript reporting his team's findings to the London office of *Nature*. His team
was now primed and ready to characterize a homologous gene (E5-1)
which they discovered while cloning S182 in January 1995. Until
now—with my exception—the contents of the discoveries had
remained confidential within a core of the research team at the
University of Toronto. Expectantly we awaited the comments of
Nature's three designated reviewers. Then, scarcely ten days later,
and before his paper had been accepted, Hyslop learned that the
mandatory confidentiality of the review process had been violated.
Hyslop was abruptly deluged with inquiries requesting confirmation that he had cloned the gene. Some callers had also heard that
the gene encoded a membrane-spanning protein with five domains.
Others had heard that the gene encoded a protease.

The ethical lapse in the review process equally outraged edi-
tors at *Nature*. The lapse was unfair both to Hyslop's team and
their competitors. For Hyslop, the lapse meant that his priority of
discovery was now jeopardized. If a competing group had heard the
almost correct version of the leak and also had a transcript characterizing a multi-spanning membrane protein on hand, then they
need test only this gene, find the mutation, and conceivably submit
a preliminary note to a journal with a publication time more rapid
than *Nature's*. And for those competitors still tenaciously searching for the gene, the fact that crucial information had leaked might
tarnish any future claim for their having independently discovered
the gene prior to the publication of Hyslop's finding.

Once Hyslop's paper received its final acceptance on June 7,
Nature accelerated its publication. Tense weeks still followed, yet
no other claims for discovery came forth. Finally, the manuscript
bearing the unassuming title "Cloning of a gene bearing missense
mutations in early-onset familial Alzheimer's disease" appeared in
the June 29, 1995 issue of *Nature*.[4]

The profound significance of the discovery was instantly recognized. Scientists foresaw that the gene's discovery would make it possible to determine whether the abnormal protein led to the disease
through a final common pathway involving the metabolism of β-amyloid—thus testing the primacy of β-amyloid models—or by entirely
different or additional mechanisms. Either way the anticipated harvest
of new results would eventually open the way for the rational treatment of early-onset Alzheimer's disease. Even though mutations in
chromosome 14 are not expected to cause late-onset Alzheimer's, an
understanding of the role of the new pivotal membrane protein and

its interactions with amyloid proteins and apoE is bound to provide insight into the mechanisms of late-onset Alzheimer's.

There would be many major discoveries secondary to the discovery of the S182 or presenilin 1 gene, but the first of these was the discovery of the highly homologous E5-1 gene. Hyslop's team mapped this novel gene to chromosome 1 and showed that it was the site of the mutation causing familial Alzheimer's disease in the Volga German families.[5] Hyslop's group named this new gene "presenilin 2." Simultaneously, Schellenberg's group, pursuing linkage studies, independently mapped the locus for familial Alzheimer's in Volga German pedigrees to chromosome 1[6], then used Hyslop's sequence for the S182 gene to find the homologous gene[7] and the same Volga German mutation.[8] Their paper, first-authorized by Ephrat Levy-Lahad and the Toronto group's paper first authorized by Evgeny Rogaev were both published in August, 1995 in *Science*[8] and *Nature*[5] respectively.

Although Hyslop's team found a second and different mutation on the chromosome 1 gene in a family from Florence studied by Sandro Sorbi and Luigi Amaducci, mutations in presenilin 2 are apt to be rare because his team failed to find this mutation in 23 other pedigrees free of mutation in other known genes.[5] The functional significance of the chromosome 1 gene and its mutations are apt to be the same as those of the chromosome 14 gene because the amino acid sequence of their two proteins are so similar and almost identical in the transmembrane domains where all deleterious mutations observed to date have been found.

If the diverse initial cell malfunctions for all forms of Alzheimer's lead to a final common metabolic pathway which could be interrupted prior to cell injury, scientists would be in the best position for finding common treatments for early- and late-onset Alzheimer's. One of the most tantalizing questions before us now is whether the overproduction of β-amyloid represents one crucial or even the crucial step in this process. Neuropathological findings in the brains of patients dying with Alzheimer's caused by mutations in the chromosome 14 gene are apparently largely indistinguishable from those with Alzheimer's caused by mutations in the amyloid gene and also from those with late-onset sporadic Alzheimer's disease.[9] All of these brains show intense depositions of β-amyloid in the same brain regions. Moreover, Steven Younkin's group showed that certain cells from patients carrying mutations in the amyloid gene secreted increased amounts of the long β-amyloid.[10] These results are apt to be extremely important because a transgenic mouse which overexpresses a mutant human amyloid gene progressively develops many of the pathological characteristics of Alzheimer's disease.[11] More recently, the laboratories of both Younkin and Dennis Selkoe have reported this increased secretion of long β-amyloid in cells from carriers of chromosome 14 mutations,[12] although it is not yet known whether this increase represents the consequence of the primary abnormality or is a secondary effect. Similarly, Hyslop's group has shown that these same mutations in presenilin lead to dramatic and accelerated accumulations of insoluble long β-amyloid in the brain itself.[13] It should not be long now before scientists can firmly establish whether the overproduction of β-amyloid, especially its longer form, invariably precedes and is a major contributor to brain cell injury or is simply a consequence of prior injury. The tools to resolve this almost century-old puzzle are finally in hand.

On April 5, 1995 Charles phoned with tragic news. After a short illness and some laboratory tests, a liver biopsy revealed a cancer that had spread to his liver presumably from his pancreas. As a pathologist, Charles knew that his life expectancy was probably 3 to 9 months. Stunned, I encouraged Charles, "Hang in there. Peter will have the gene very soon." "It had better be darn soon," Charles wryly responded.

Charles said that there was another reason for his call. The Chedd-Angier Production Company had been planning a proposed TV documentary for PBS based on his family. Charles, Ben, and Elizabeth had agreed to participate. Independently, each of them had decided to waive confidentiality with respect to their contribution. Each believed that an in-person appearance would have even greater impact on the story and gain increasing support for research on Alzheimer's disease than an "in shadow," disguised appearance. Charles asked me to contact Chedd-Angier and convey his stark assessment, "They had better come right away." A week-and-a-half later the TV crew arrived and Charles, between occasional rests, told his story during lengthy filmed sessions.

I also told Peter Hyslop of Charles's grave illness. "I'll call," Peter responded with deep sadness and respect. I believe that Peter's advance notice of the gene's discovery helped sustain Charles during the following months. He eagerly awaited the publication of the paper and yearned to read of it in his own copy of *Nature*.

On June 28th, as the announcement of the gene discovery

neared, *USA Today* ran its front-page feature news story on the discovery and its accompanying cover story based on an interview with Ben, Dr. Ben Williamson of Rock Island, Illinois, who acknowledged with pride the role that he and Dr. Charles Altshuler of Milwaukee, Wisconsin, had played in the long struggle. That evening ABC's *World News Tonight* carried the news story and included comments from "Elizabeth," Mrs. Betty Weiss of Cherry Hill, New Jersey. The next morning, ABC's *Good Morning America* ran a sensitive and informative account of the history of Charles's family and the gene's discovery. Charles appeared "in shadow" from footage taped months earlier.

Another event brought great joy to Charles. His grand-niece Laura, the daughter of my patient Jeff, had decided after the discovery of the gene linkage that she and her husband would risk having a child. She reasoned that even in the worse case that both she and her child had inherited Jeff's gene that there would be treatment 40 years hence. Charles had lived to learn of the birth of his great-grandnephew in early July.

As July progressed, Charles's strength ebbed mercilessly. The thought obsessed him that his records and collection of tissues might well be lost. By the third week of July, he sensed that his last chance had come. Only later I learned that he enlisted the help of his long-time professional colleague Dr. John Bareta, and returned to his research office at Milwaukee's St. Joseph's Hospital one last time. He identified each treasured record and extracted a promise from Bareta to "send them to Hyslop so that they might help someone someday."

The time may come when I may not recall that it was a change in only a single amino acid from cysteine to tyrosine at residue number 410 in the S182 protein that caused Alzheimer's disease in Hannah's family, but as long as memory endures I shall not forget her courageous family and especially Dr. Charles Altshuler who near the end, like Tennyson's Ulysses, still strove "to seek, to find, and not to yield." I shall surely miss him.

In Remembrance
Dr. Charles H. Altshuler
April 14, 1919–July 29, 1995
He stood—at first alone—against the flood, and held.

February 15, 1996

Dr. Ben Williamson. (Courtesy of Dr. Williamson.)

Betty Weiss. (Courtesy of Mrs. Weiss.)

Dr. Charles Altshuler. (Courtesy of the Chedd-Angier Production Company.)

Dr. Dennis Selkoe. (Courtesy of Dr. Selkoe.)

"Hyslop's team" from left to right are: Drs. Peter St. George-Hyslop, Evgeny Rogaev, Paul E. Fraser, Robin Sherrington, and Johanna Rommens (seated). (Courtesy of Rick Chard.)

Notes

The Life and Death of Hannah

1. Details of the lives of Shlomo and Hannah were provided by Ben W. and Bryna M., respectively.

2. For details of the migration of the Jews into Eastern Europe and of their lives there, see S. M. Dubnow and I. Friedlander, *The History of the Jews in Russia and Poland*, Vols. I, II (Philadelphia: Jewish Publication Society of America, 1918).

3. For some details on the history of Bobruisk and Byelorussia, see *Great Soviet Encyclopedia* (New York: Macmillan, 1983).

Gregor Mendel

1. Vitezslav Orel, *Gregor Mendel* (New York: Oxford University Press, 1984). Provides a history of Mendel's life and work and documents the scientific tradition to which Mendel was heir.

2. Orel, pp. 9–17.

3. *Experiments in Plant Hybridization*, Mendel's original papers in

English translation, with commentary and assessment by Ronald Fisher, edited by J. H. Bennett (London: Oliver and Boyd, 1965). This work contains a translation of Mendel's paper and insights into Mendel's life.

4. Sam Singer, *Human Genetics: An Introduction to the Principle of Heredity* (New York: W. H. Freeman and Co., 1978), p. 139. (See especially pp. 1–7.) This work provides a fine introduction to genetics suitable for the general reader.

5. Benjamin Lewin, *Genes IV*, (New York: Oxford University Press, 1991). Provides a detailed text for the serious student of molecular genetics.

6. See Harry Sootin, *Gregor Mendel, Father of the Science of Genetics* (New York: Vanguard, 1959).

Alois Alzheimer

1. F. H. Lewey, *Alois Alzheimer*, in *The Founders in Neurology* compiled and edited by Webb Haymaker and Frances Schiller, 2nd ed. (Springfield, Ill.: Charles C. Thomas, 1970), pp. 315–318.

2. H. Lauter, "Some Remarks on the Second Medical Faculty in Munich and on the Life and Scientific Achievements of Alois Alzheimer," Lecture given at a Psychiatric Symposium in Munich, October 4, 1990. See also the detailed written history by Lauter's two colleagues, Drs. Hoff and Hippius: P. Hoff and H. Hippius, "Alzheimer Alois, 1864–1915. Ein Uberblick uber Leben und Werk anlβslich seines 125," *Geburstages, Nervenarzt* 60 (1989): 332–337.

3. Arthur W. Young "Franz Nissl and Alois Alzheimer," *Archives of Neurology and Psychiatry* 33 (1935): 847–852.

4. Lewey, p. 317.

5. Lauter.

6. Lewey, pp. 315–316.

7. Lewey, p. 315.

8. Lewey, p. 317.

9. Thanks to an extraordinary undertaking by Katherine Bick with the support of the Italian study group in Brain Aging (ISGBA), the early papers by Alois Alzheimer (1906), Oskar Fischer, Francesco Bonfiglio, Emil Kraepelin, and Gaetano Persini can now be read in English translation. See Katherine Bick, Luigi Amaducci, and Giancarlo Pepeu, eds., *The Early Story of Alzheimer's Disease* (Padua, Italy: Lavonia Press, 1987. New York: Raven Press, 1987). The titles in English translation and in their original languages are:

Alois Alzheimer, *A Characteristic Disease of the Cerebral Cortex* (Original title: Über eine eigenartige Erkrankung der Hirnrinde)

Oskar Fischer, *Miliary Necrosis with Nodular Proliferation of the Neurofibrils, a Common Change of the Cerebral Cortex in Senile Demen-*

tia (Original title: Miliare Nekrosen mit drusigen Wucherungen der Neurofibrillen, eine regelmässige Veränderung der Hirnrinde bei seniler Demenz)

Francesco Bonfiglio, *Concerning Special Findings in a Case of Probable Cerebral Syphilis* (Original title: Di speciali reperti in un caso di probabile sifilide cerebrale)

Emil Kraepelin, *Senile and Pre-Senile Dementias* (Original title: Das senile und prasenile Irresein)

Gaetano Perusini, *Histology and Clinical Findings of Some Psychiatric Diseases of Older People* (Original title: Über klinisch und histologisch eigenartige psychische Erkrankungen des späteren Lebensalters)

Gaetano Perusini, *The Nosographic Value of Some Characteristic Histopathological Findings in Senility* (Original title: Sul valore nosografico di alcuni reperti istopatologici caratteristici per la senilità)

10. Alzheimer in Bick, Amaducci, and Pepau, p. 2.
11. The first English translation of important segments of Alois Alzheimer's 1911 paper on Alzheimer's disease, with commentary, was published in 1988. See Joseph Tonkonogy and Gary Moak, "Alois Alzheimer on Presenile Dementia," *Journal of Geriatric Psychiatry in Neurology*, 1 (1988): 199–206.
12. Fischer in Bick, Amaducci, and Pepeu.
13. Bonfiglio in Bick, Amaducci, and Pepau.
14. Kraepelin in Bick, Amaducci, and Pepau.
15. Tonkonogy and Moak, p.
16. Perusini in Bick, Amaducci, and Pepeu.
17. I am especially grateful to Dr. Hans-Dieter Lux of the Max Planck Institute for Psychiatry in Munich for finding the obituary on the life of Alois Alzheimer by Robert Gaupp from the *Muenchener Medizinische Wochenschrift*, February 1916, pp. 195–196, and to Dr. Ursula Anwer for translating the paper into English.
18. Gaupp, p. 196.

Mendel's Heirs

1. Ian Shine and Sylvia Wrobel, *Thomas Hunt Morgan* (Lexington, Kentucky: University of Kentucky Press, 1976; and Garland E. Allen, *Thomas Hunt Morgan: The Man and His Science* (Princeton, N.J.: Princeton University Press, 1978).
2. Shine and Wrobel, p. 56.
3. Shine and Wrobel, p. 56.
4. Shine and Wrobel, pp. 73–74.

5. Shine and Wrobel, p. 74.

6. Shine and Wrobel, pp. 74–75.

7. Shine and Wrobel, p. 92.

8. Allen, p. 389.

9. Horace Freeland Judson, *The Eighth Day of Creation* (New York: Simon and Schuster, 1979), p. 50.

10. Judson, p. 47.

11. L. Pauling and M. Delbrück, "The Nature of the Intermolecular Forces Operative in Biological Processes," *Science* 92 (1940): 77–79.

12. James D. Watson, John Tooze, and David Kurtz, *Recombinant DNA: A Short Course* (New York: Scientific American Books, 1983).

13. Watson, Tooze, and Kurtz, pp. 11–12.

14. Watson, Tooze, and Kurtz, p. 12.

15. Judson, p. 88.

16. Judson, pp. 109–110.

17. Judson, p. 173.

18. Judson, p. 96.

19. Judson, p. 178.

20. J. D. Watson and F. H. C. Crick, "Molecular Structure of Nucleic Acids: A Structure for Deoxyribose Nucleic Acid," *Nature* 171 (1953): 737–738.

Alzheimer's Disease at Midcentury

1. R. D. Newton, "The Identity of Alzheimer's Disease and Senile Dementia and Their Relationship to Senility," *Journal of Mental Science* 94 (1948): 225–249. (Please refer to Newton's bibliography for references to the early work cited in his paper.)

2. W. Mayer-Gross, "Recent Progress in Psychiatry, Arteriosclerotic, Senile and Presenile Dementia," *Journal of Mental Science* 90 (1944): 316–327.

3. F. Struwe, "Histopathologische Untersuchungen Uber Entstehung Und Wesen Der Senilen Plaques," *Zentralblatt fur Nervenheilkunde und Psychiatrie* 122 (1929): 291–307.

4. G. A. Jervis, "Early Senile Dementia in Mongoloid Idiocy," *American Journal of Psychiatry* 105 (1948): 102–106.

Enter Genetics

1. The early history of the blood groups in humans can be found in R. R. Race and Ruth Sanger, *Blood Groups in Man*, 5th ed. (Philadelphia: F. A. Davis Co., 1968).

2. Race and Sanger, p. 1.

3. Race and Sanger, p. 1.

4. Race and Sanger, p. 1.

5. C. A. B. Smith, "The Detection of Linkage in Human Genetics," *Journal of the Royal Statistical Society* 15 (1953): 153–192; and N. E. Morton, "Further Scoring Types in Sequential Linkage Tests, With a Critical Review of Autosomal and Partial Sex Linkage in Man," *American Journal of Human Genetics* 9 (1957): 55–75.

6. N. Risch, "Genetic Linkage: Interpreting Lod Scores," *Science* 255 (1992): 803–804. There is a non-zero a priori chance that any two markers or a marker and a disease will be linked. The a posteriori odds for linkage between two markers are lower than the lod scores as they are conventionally calculated by a factor of about fifty.

7. Other work on the blood groups and the concept of linkage in general is reviewed, with the primary references given in V. A. McKusick and Frank H. Ruddle, "The Status of the Gene Map of the Human Chromosomes," *Science* 196 (1977): 390–405.

8. We know today that the genes for the MN system are located on the long arm of chromosome 4.

9. L. Wheelan and R. R. Race, "Familial Alzheimer's Disease," *Annals of Human Genetics* 23 (1959): 300–310.

10. Wheelan and Race, p. 301.

New Seeds Are Sown

1. Patrick Fox, "From Senility to Alzheimer's Disease: The Rise of the Alzheimer's Disease Movement," *The Milbank Quarterly* 67 No. 1 (1989): 58–102.

2. Fox, p. 66.

3. R. Terry, N. K. Gonatas, and M. Weiss, "Ultrastructural Studies in Alzheimer's Presenile Dementia," *American Journal of Pathology* 44 (1964): 269–297.

4. Fox, p. 67.

5. R. Katzman, K. Suzuki and S. Korey, "Chemical Studies on Alzheimer's Diease," *Neuropathology and Experimental Neurology* 24 (1965): 211–224.

6. Fox, p. 70.

7. W. J. McEntel and T. H. Crook, "Age-Associated Memory Impairment: A Role for Catecholamines," *Neurology* 40 (1990): 526–530.

No Longer Alone

1. G. E. W. Wolstenholme and M. O'Connor, *Ciba Symposium on Alzheimer's Disease and Related Conditions* (London: Churchill, 1979), p. 360.

2. R. Katzman, "The Prevalence and Malignancy of Alzheimer's Disease," *Archives of Neurology* 33 (1976): 217–218.

3. R. Katzman and T. Karasu, "Differential Diagnosis of Dementia," in W. S. Fields, ed., *Neurological and Sensory Disorders in the Elderly* (New York: Stratton Intercontinental Medical Book Corp., 1975), pp. 103–134, 106.

4. Katzman, *Archives of Neurology*, p. 218.

Twin Pillars of Hope

1. Patrick Fox, "From Senility to Alzheimer's Disease: The Rise of the Alzheimer's Disease Movement," *The Milbank Quarterly* 67, No. 1 (1987): 58–102.

2. Details on the contributions of Florence Mahoney were provided by Jerome Stone and Drs. Marrott Sinex and Robert Butler.

3. Fox, p. 79.

4. Bobbie Glaze Custer, *I Chose to Continue* (Minneapolis: Lakeland Color Press, 1991).

5. Custer, personal communication.

6. Fox, p. 85.

7. Robert Butler, personal communication.

8. Robert Katzman, Robert D. Terry, and Katherine L. Bick, "Recommendations of the Nosology, Epidemiology, and Etiology and Pathophysiology Commissions of the Workshop-Conference on Alzheimer's Disease, Senile Dementia and Related Disorders," in *Alzheimer's Disease: Senile Dementia and Related Disorders* (Aging, Vol. 7), ed. by R. Katzman, R. D. Terry, and K. L. Bick (New York: Raven Press, 1978), pp. 579–585.

A Transmissible Virus?

1. D. C. Gajdusek and V. Zigas, "Kuru," *American Journal of Medicine* 26 (1959): 442–469.

2. D. C. Gajdusek, "Unconventional Viruses and the Origin and Disappearance of Kuru," *Science* 197 (1977): 943–960.

3. See reviews by S. B. Prusiner, "Prions and Neurodegenerative Diseases," *New England Journal of Medicine* 317 (1987): 1571–1581, and by P. Brown, L. G. Goldfarb, and D. C. Gajdusek, "The New Biology of Spongiform Encephalopathy: Infectious Amyloidosis with a Genetic Twist," *Lancet* 337 (1991): 1019–1022, and by S. B. Prusiner, "Molecular Biology of Prion Proteins," *Science* 252 (1991): 1515–1522.

4. P. Brown, "Central Nervous System Amyloidoses: A Comparison of

Alzheimer's Disease and Creutzfeldt-Jakob Disease," *Neurology* 39 (1989): 1103–1105.

5. J. Cherfas, "Mad Cow Disease: Uncertainty Rules," *Science* 249 (1990): 1492–1493.

Base Camp, Circa 1980

1. W. Arber, "Promotion and Limitation of Genetic Exchange," *Science* 205 (1979): 361–365.

2. H. O. Smith, "Nucleotides Sequence Specificity of Restriction Endonucleases," *Science* 205 (1979): 455–462.

3. D. Nathans, "Restriction Endonucleases, Simian Virus 40, and the New Genetics," *Science* 206 (1979): 903–909.

4. Smith, p. 455.

5. Recombinant DNA technology refers to that set of molecular techniques whereby select strips of DNA can be clipped into desired fragments, fused with other specified fragments, and thence injected back into the cells of an animal or plant of the same or a different species so as to amplify the production of naturally occurring proteins or to generate new gene products.

A New Basis for Linkage Maps of the Human Genome

1. A complex of globin genes directs the synthesis of the protein portion of hemoglobin, the carrier of oxygen in red blood cells.

2. Y. Kan and A. Dozy, "Antenatal Diagnosis of Sickle-cell Anaemia by DNA Analysis of Amniotic-Fluid Cells," *Lancet* 2 (1978): 910–912. Y. W. Kan and A. M. Dozy, "Polymorphisms of DNA Sequence Adjacent to Human β-globin Structural Gene: Relationship to Sickle Mutation," *Proceedings of the National Academy of Science* 75 (1978): 5631–5635.

3. E. Solomon and W. F. Bodmer, "Evolution of Sickle Variant Gene," *Lancet* 1 (1979): 923.

4. Crossovers occur at varying rates across the genome, and the recombination rate at different genetic loci may vary greatly depending on the sex of the parent. Even so, the figures used here serve as reasonable values for many regions.

5. M.-C. King and A. C. Wilson, "Evolution at Two Levels in Humans and Chimpanzees," *Science* 188 (1975): 107–116.

6. A. J. Jeffreys, "DNA Sequence Variants in the $^G\gamma$-, $^A\gamma$-, δ- and β-globin Genes of Man," *Cell* 18 (1979): 1–10.

7. Jeffreys, p. 9.

8. Jerry E. Bishop and Michael Waldholz, *Genome* (New York: Simon and Schuster, 1990).

9. Bishop and Waldholz, pp. 60–61.

10. D. Bostein, R. L. White, M. Skolnick, and R. W. Davis, "Construction of a Genetic Linkage Map in Man Using Restriction Fragment Length Polymorphisms," *American Journal of Human Genetics* 32 (1980): 314–331.

11. Botstein et al., p. 317.

12. Botstein et al., pp. 317–318.

13. Botstein et al., pp. 314–331.

14. A. R. Wyman and R. L. White, "Construction of a Genetic Linkage Map in Man Using Restriction Fragment Length Polymorphisms," *American Journal of Human Genetics* 32 (1980): 314–331.

The Quest for the Huntington's Disease Gene

1. Nancy Wexler, "Life in the Lab," *Los Angeles Times Magazine*, February 10, 1991, pp. 12–13 and 31.

2. Jerry E. Bishop and Michael Waldholz, *Genome* (New York: Simon and Schuster, 1990). I have relied on this well-written and well-researched book for the history of the Wexler family, their commitment to research on Huntington's disease, the founding of the Hereditary Disease Foundation by Milton Wexler, and the early history of Huntington's disease research in Venezuela.

3. Bishop and Waldholz, p. 30.

4. Wexler, p. 12.

5. Bishop and Waldholz, p. 37.

6. Bishop and Waldholz, p. 42.

7. From Nancy Wexler's curriculum vitae and personal communication.

8. Bishop and Waldholz, p. 44.

9. It is still not known whether Huntington's disease arose as a *de novo* mutation in this native American community or was introduced by European sailors known to have reached Lake Maracaibo during the eighteenth century.

10. Bishop and Waldholz, p. 91.

11. From an as yet unpublished manuscript by Nancy Wexler, cited in Bishop and Waldholz, p. 98.

Charles

1. References to some of the major scientific studies on familial Alzheimer's disease made possible in part by the cooperation of the descendants of Hannah and Shlomo in the period prior to the age of modern molecular genetics are summarized below.

R. D. Terry, N. K. Gonatas, and M. Weiss, "Ultrastructural Studies in Alzheimer's Presenile Dementia," *American Journal of Pathology* 44 (1964): 269–277.

R. D. Terry, "Electron Microscopic Studies of Alzheimer's Disease and of Experimental Neurofibrillary Tangles," In *The Central Nervous System*, International Academy of Pathology Monograph No. 9. Baltimore: Williams and Wilkins Co., 1968, Chapter 12, pp. 213–224.

K. Suzuki, R. Katzman, and S. R. Korey, "Chemical Studies on Alzheimer's Disease," *Journal of Neuropathology and Experimental Neurology* 24 (1965): 211–224.

J. Goudsmit, B. J. White, L. R. Weitkamp, B. J. B. Keats, C. H. Morrow, and D. C. Gadjusek, "Familial Alzheimer's Disease in Two Kindreds of the Same Geographic and Ethnic Origin," *Journal of Neurological Science* 49 (1981): 79–89.

J. Goudsmit, C. H. Morrow, D. M. Asher, R. T. Yanagihara, C. L. Masters, C. J. Gibbs, Jr. and D. C. Gadjusek, "Evidence for and against the Transmissibility of Alzheimer's Disease," *Neurology* 30 (1980): 945–950.

Peter St. George-Hyslop and the Link to Chromosome 21

1. Dr. David Drachman is chairman of the Department of Neurology at the University of Masachusetts Medical Center. He first worked with Robert Katzman and later succeeded him as chairman of the Medical and Scientific Advisory Board of the Alzheimer's Disease and Related Disorders Association. Drachman also initially proposed and defined the need for, and the design of, NIH's now implemented Alzheimer's Disease Research Center program, which was made possible by ADRDA's political leverage and a receptive U.S. Congress. Having completed his term as chairman of a board of ADRDA, Drachman remains a scientific and medical spokesman for the Alzheimer movement, as well as an advocate for caregivers and patients.

2. G. G. Glenner and C. W. Wong,'Alzheimer's Disease and Down's Syndrome: Sharing of a Unique Cerebrovascular Amyloid Fibril Protein," *Biochemical and Biophysical Research Communications* 122 (1984): 1131–1335.

3. L. E. Nee, R. J. Polinsky, R. Eldridge, H. Weingartner, S. Smallberg, and M. Ebert, "A Family with Histologically Confirmed Alzheimer's Disease," *Archives of Neurology* 40 (1983): 203–209.

4. R. G. Feldman, K. A. Chandler, L. L. Levy, and G. H. Glazer, "Familial Alzheimer's Disease," *Neurology* 13 (1963): 811–824.

5. J. F. Foncin, D. Salmon, V. Supino-Viterbo, R. G. Feldman, G. Macchi, P. Mariotti, C. Scopetta, G. Caruso, and A. C. Bruni, "Demence Presenile d'Alzheimer Transmise Dans Une Famille Etendue," *Revue Neurologic* (Paris) 141 (1985): 194–202.

6. P. H. St. George-Hyslop, R. E. Tangi, R. J. Polinsky, J. L. Haines, L. Nee, P. C. Watkins, R. H. Myers, R. G. Feldman, D. Pollen, D. Drachman, J. Growdon, A. Bruni, J.-F. Foncin, D. Salmon, P. Frommelt, L. Amaducci, S. Sorbi, S. Piacentini, G. D. Stewart, W. J. Hobbs, P. M. Conneally, and J. F. Gusella, "The Genetic Defect Causing Familial Alzheimer's Disease Maps on Chromosome 21," *Science* 235 (1987): 885–890.

A Gene for Amyloid

1. D. Goldgaber, M. I. Lerman, O. W. McBride, U. Saffiotti, and D. C. Gajdusek, "Characterization and Chromosomal Localization of a cDNA Encoding Brain Amyloid of Alzheimer's Disease," *Science* 235 (1987): 877–880.

2. R. E. Tanzi, J. F. Gusella, P. C. Watkins, G. A. P. Bruns, P. St. George-Hyslop, M. L. Van Keuren, D. Patterson, S. Pagan, D. M. Kurnit, and R. L. Neve, "Amyloid β-protein Gene: cDNA, mRNA Distribution and Genetic Linkage near the Alzheimer's Locus," *Science* 235 (1987): 880–884.

3. J. Kang, H.-G. Lemaire, A. Unterbeck, J. M. Salbaum, C. L. Masters, K.-H. Grzeschik, G. Multhaup, K. Beyreuther, and B. Müller-Hill, "The Precursor of Alzheimer's Disease Amyloid A4 Protein Resembles a Cell-Surface Receptor," *Nature* 325 (1987): 733–736.

4. N. K. Robakis, H. M. Wisniewski, E. C. Jenkins, E. A. Devine-Gage, G. E. Houck, X.-L. Yao, N. Ramakrishna, G. Wolfe, W. P. Silverman, and W. T. Brown, "Chromosome 21q21 Sublocalisation of Gene Encoding Beta-Amyloid Peptide in Cerebral Vessels and Neuritic (Senile) Plaques of People with Alzheimer's Disease and Down Syndrome," *Lancet* (February 11, 1987): 384–385.

Russia Once Again

1. C. Holden, "Politics and Soviet Psychiatry," *Science* 239 (1988): 551–553.

The Twenty-One Pedigrees of Nina Ivanovna Voskresenskaya

1. I am indebted to Dr. Svetlana Ivanovna Gavrilova for sending me autobiographical details on the life of Erich Sternberg.

How Many Genes?

1. F. C. Koch, *The Volga Germans* (University Park: Pennsylvania State University Press, 1977), p. 340.

2. T. D. Bird, T. H. Lampe, E. K. Nemens, G. W. Miner, S. M. Sume, and G. D. Schellenberg, "Familial Alzheimer's Disease in American Descendants of the Volga Germans: Probable Genetic Founder Effect," *Annals of Neurology* 25 (1989): 12–25.

The Third Dove Goes Forth

1. C. L. Joachim, H. Mori, and D. J. Selkoe, "Amyloid β-protein Deposition in Tissues Other Than Brain in Alzheimer's Disease," *Nature* 341 (1989): 226–230.

2. Selkoe's hypothesis that deposition of amyloid in the skin would distinguish Alzheimer's victims from normal, middle-aged individuals was not confirmed by later work. Over the next two years, Selkoe's group studied skin samples from both Alzheimer's victims and escapees in Charles's and in Elizabeth's families, and could not predict on the basis of the presence or absence of amyloid deposition in the skin whether or not the donor had Alzheimer's disease or was normal. Selkoe reported these negative findings at a meeting of the American Neurological Association in Boston in April 1991 and full publication of these results is pending. However, the issues of whether amyloid deposits outside the brain are related to the pathogenesis of Alzheimer's disease and whether amyloid depositions within the brain are toxic to nerve cells remain highly controversial. The September 4, 1992 issue of *Science* carried a commentary on these controversies. J. Marx, "Alzheimer's Debate Boils Over," *Science* 257 (1992): 1336–1338.

Lost Tribes

1. K. B. Mullis, "The Unusual Origin of the Polymerase Chain Reaction," *Scientific American* 262 (1990): 56–65. See also commentary by J. L. Marx, "Multiplying Genes by Leaps and Bounds," *Science* 240 (1988): 1408–1410.

The First Abnormal Gene Is Found

1. E. Levy, M. D. Carman, I. J. Fernandez-Madrid, M. D. Power, I. Leiberburg, S. G. Van Duinen, G. T. A. M. Bots, W. Luydendijk, and B. Frangione, "Mutation of the Alzheimer's Disease Amyloid Gene in Hereditary Cerebral Hemorrhage, Dutch Type," *Science* 248 (1990): 1124–1126.

2. J. Haan, J. B. K. Lanser, I. Zijderveld, I. G. F. van der Does, and R. A. C. Roos, "Dementia in Hereditary Cerebral Hemorrhage with Amyloidosis-Dutch Type," *Archives of Neurology* 47 (1990): 965–968.

3. A. Goate, M.C. Chartier-Harlin, M. Mullan, J. Brown, F. Crawford, L. Fidani, L. Giuffra, A. Haynes, N. Irving, L. James, R. Mant, P. Newton, K. Rooke, P. Roques, C. Talbot, M. Pericak-Vance, A. Roses, R. Williamson, M. Rossor, M. Owen, and J. Hardy, "Segregation of a Missense Mutation in the Amyloid Precursor Protein Gene with Familial Alzheimer's Disease," *Nature* 349 (1991): 704–706.

4. "ISI's Hot Papers Database, 1991," *Science Watch*, January 1992. This source is on a computerized database and not in a journal. However, *Nature* has listed these frequently cited papers in an ad opposite page 813 in *Nature* 356 (1992).

5. J. Murrell, M. Farlow, B. Ghetti, and M. D. Benson, "A Mutation in the Amyloid Precursor Protein Associated with Hereditary Alzheimer's Disease," *Science* 254 (1991): 97–99.

6. M.-C. Chartier-Harlin, F. Crawford, H. Houlden, A. Warren, D. Hughes, L. Fidani, A. Goate, M. Rossor, P. Roques, J. Hardy, and M. Mullan, "Earlier-onset Alzheimer's Disease Caused by Mutations at Codon 717 of the β-amyloid Precursor Protein Gene," *Nature* 353 (1991): 844–846.

7. J. Hardy and D. Allsop, "Amyloid Deposition as the Central Event in the Aetiology of Alzheimer's Disease," *Trends in Pharmacological Science* 12 (1991): 383–388 and J. A. Hardy and G. A. Higgins, "The Amyloid Cascade Hypothesis," *Science* 256 (1992): 184–185.

8. J. Marx, "Alzheimer's Debate Boils Over," *Science* 257 (1992): 1336–1338.

9. P. Coleman, ed. September issue of *Neurobiology of Aging* 13 (1992): No. 5.

10. S. Kawabata, G. A. Higgins, and J. W. Gordon, "Amyloid Plaques, Neurofibrillary Tangles and Neuronal Loss in Brains of Transgenic Mice Overexpressing a C-terminal Fragment of Human Amyloid Precursor Protein," *Nature* 354 (1991): 476–478. See also retraction of this paper in *Nature* 356 (1992): 23, and in *Nature* 356 (1992): 265.

11. J. Rennie, "The Mice That Missed," *Scientific American* 266 (1992): 20–21.

New Sightings

1. M. Mullan, F. Crawford, K. Axelman, H. Houlden, L. Lilius, B. Winblad, and L. Lannfelt, "A Pathogenic Mutation for Probable Alzheimer's Disease in the APP Gene at the N-Terminus of β-amyloid," *Nature Genetics* 1 (1992): 345–347.

2. R. E. Tanzi, G. Vaula, D. M. Romano, M. Mortilla, T. L. Huang, R. G. Tupler, W. Wasco, B. T. Hyman, J. L. Haines, B. J. Jenkins, M. Kalaitsidaki, A. C. Warren, M. C. McInnis, S. E. Antonarakis, H. Karlinsky, M. E.

Percy, L. Connor, J. Growdon, D. R. Crapper-McLachlan, J. F. Gusella, and P. H. St. George-Hyslop, "Assessment of Amyloid β-Protein Precursor Gene Mutations in a Large Set of Familial and Sporadic Alzheimer Disease Cases," *American Journal of Human Genetics* 51 (1992): 273–282.

3. M. A. Pericak-Vance, J. L. Bebout, P. C. Gaskell, Jr., L. H. Yamaoka, W. Y. Hung, M. J. Alberts, A. P. Walker, R. J. Bartlett, C. A. Haynes, K. A. Welsh, N. L. Earl, A. Heyman, C. M. Clark, and A. D. Roses, "Linkage Studies in Familial Alzheimer's Disease: Evidence for Chromosome 19 Linkage," *American Journal of Human Genetics* 48 (1991): 1034–1050.

4. R. M. Cook-Deegan in his forthcoming book *The Gene Wars: Science, Politics and the Human Genome* (New York: W. W. Norton, 1993) traces the roots, evolution, and current status of the human genome project.

5. E. Jordan, "Invited Editorial: The Human Genome Project: Where Did It Come From, Where Is It Going?" *American Journal of Human Genetics* 51 (1992): 1–6.

6. J. D. Watson, "The Human Genome Project: Past, Present and Future," *Science* 248 (1990): 44–49; quote on p. 44.

7. Watson, p. 46.

8. Watson, p. 46.

Mutual Aid as a Factor in Evolution and in Genetic Research

1. P. A. Kropotkin, *Mutual Aid: A Factor of Evolution* (New York: McClure Phillips and Co., 1902), p. *ix*.

2. Kropotkin, p. *x*.

3. Kropotkin, p. 2.

4. E. O. Wilson, *Sociobiology: The New Synthesis* (Cambridge, Mass.: The Belnap Press of Harvard University, 1975), p. 697.

5. Wilson, p. 593.

6. Wilson, pp. 3–4.

7. L. Roberts, "The Race for the Cystic Fibrosis Gene," *Science* 240 (1988): 141–144.

8. M. Conneally, "A First Step Toward a Molecular Genetic Analysis of Amyotrophic Lateral Sclerosis," *New England Journal of Medicine* 324 (1991): 1430–1432.

9. A. D. Roses and P. M. Conneally, "Alzheimer's Disease Cell Bank," *Science* 252 (1991): 14.

10. E. Marshall, "Data Sharing: A Declining Ethic?" *Science* 248 (1990): 952–957.

Epilogue: Landfall

1. G. D. Schellenberg, T. D. Bird, E. M. Wijsman, H. T. Orr, L. Anderson, E. Nemens, J. A. White, L. Bonnycastle, J. L. Weber, M. E. Alonso, H.

Potter, L. L. Heston, and G. M. Martin, "Genetic Linkage Evidence for a Familial Alzheimer's Disease Locus on Chromosome 14," *Science* 258 (1992): 668–671.

2. P. H. St. George-Hyslop, J. Haines, E. Rogaev, M. Mortilla, G. Vaula, M. Pericak-Vance, J.-F. Foncin, M. Montesi, A. Bruni, S. Sorbi, I. Rainero, L. Pinessi, D. Pollen, R. Polinsky, J. Nee, J. Kennedy, F. Macciardi, E. Rogaeva, Y. Liang, N. Alexandrova, W. Lukiw, K. Schlumpf, R. Tanzi, T. Tsuda, L. Farrer, J-M Cantu, R. Duara, L. Amaducci, L. Bergamini, J. Gusella, A. Roses, and D. C. McLachlan, "Genetic Evidence for a Novel Familial Alzheimer's Disease Locus on Chromosome 14," *Nature Genetics* 2 (1992): 330–334.

3. C. Van Broeckhoven, H. Backhovens, M. Cruts, G. DeWinter, M. Bruyland, P. Cras, and J-J Martin, "Mapping of a Gene Predisposing to Early-Onset Alzheimer's Disease to Chromosome 14q24.3," *Nature Genetics* 2 (1992): 335–339.

4. J. J. Martin, J. Gheuens, M. Bruyland, P. Cras, A. Vandenberghe, L. Masters, K. Beyreuther, R. Dom, C. Ceuterick, U. Lübke, H. Van Heuverswijn, G. DeWinter, and C. Van Broeckhoven, "Early-Onset Alzheimer's Disease in 2 Large Belgian Families," *Neurology* 41 (1991): 62–68.

5. L. R. Weitkamp, L. Nee, B. Keats, R. J. Polinsky, and S. Guttormsen, "Alzheimer Disease: Evidence for Susceptibility Loci on Chromosomes 6 and 14," *American Journal of Human Genetics* 35 (1983): 443–453.

6. A lod score of 3.0 signifies that the combined *a priori* and *a posteriori* odds of linkage between a marker and a genetic disease are 1,000 to 1. However, there is an *a priori* probability of linkage of about 1 in 50 between any two randomly tested markers or between a marker and a genetic disease. Hence, the *a posteriori* odds for linkage—given a lod score of 3—must be reduced some 50-fold, yielding an *a posteriori* probability of linkage of 20 to 1. For a detailed discussion of lod scores see N. Risch, "Genetic Linkage: Interpretating Lod Scores," *Science* 225(1992): 803–804.

7. Early studies following up on this result on the same family using a nearby and more informative marker called D14S1 showed negative evidence for linkage. See P. H. St. George-Hyslop, A. Polinsky, L. Nee, J. Haines, R. Tanzi, P. Conneally, J. Growdon, R. H. Myers, D. Pollen, D. Drachman, R. Feldman, L. Amaducci, J-F. Foncin, A. Bruni, P. Frommet, and J. F. Gusella, "Search for the Familial Alzheimer's Disease Gene," *Journal of Neural Transmission* 24 (1987): 13–21.

8. C. R. Abraham, D. J. Selkoe, and H. Potter, "Immunochemical Identification of the Serine Protease Inhibitor α_1-Antichymotrypsin in the Brain Amyloid Deposits of Alzheimer's Disease," *Cell* 52 (1988): 487–501.

9. J. Weber and P. E. May, "Abundant Class of Human DNA Polymorphisms Which Can Be Typed Using the Polymerase Chain Reaction," *American Journal of Human Genetics* 44 (1989): 388–396.

10. C. Hooper, "Of Rumors, Reviews, and Rip-Offs in Alzheimer's

Research," *The Journal of National Institutes of Health Research* 4 (1992): 34.

11. V. Sharma, L. Smith, L. Allen, R. E. Magenis, and M. Litt, "Dinucleotide Repeat Polymorphism at the D14S43 Locus," *Nucleic Acids Research* 19 (1991): 1722.

12. Z. Wang and J. L. Wcbcr, "Continuous Linkage Map of Human Chromosome 14 Short Tandem Repeat Polymorphisms," *Genomics* 13 (1992): 532–536.

13. Hooper, p. 34.

14. J. E. Hamos, B. Oblas, D. Pulaski-Salo, W. J. Welch, D. G. Bole, and D. A. Drachman, "Expression of Heat Shock Proteins in Alzheimer's Disease," *Neurology* 41 (1991): 345–350.

15. S. Wiggins, P. Whyte, M. Huggins, S. Adam, J. Theilmann, M. Bloch, S. B. Sheps, M. T. Schechter, and M. R. Hayden, "The Psychological Consequences of Predictive Testing for Huntington's Disease." *The New England Journal of Medicine* 327 (1992): 1401–1405.

16. J. B. Lawrence, "A Fluorescence In Situ Hybridization Approach for Gene Mapping and the Study of Nuclear Organization," *Genome Analysis Volume 1: Genetic and Physical Mapping* (1990): 1–38. Cold Spring Harbor Laboratory Press.

17. Unfortunately, the technique cannot pick up a single-base substitution, which is still the most frequent type of mutation causing genetic defects.

18. W. J. Strittmatter, A. M. Saunders, D. Schmechel, M. Pericak-Vance, J. Enghild, G. S. Salvesen, and A. D. Roses, "Apolipoprotein E: High Affinity Binding to β-Amyloid and Increased Frequency of Type 4 Allele in Familial Alzheimer's Disease," *Proceedings of the National Academy of Science* 90(1993): 1977–1981.

19. E. I. Rogaev, W. J. Lukiw, G. Vaula, J. L. Haines, E. A. Rogaeva, T. Tsuda, N. Alexandrova, Y. Liang, M. Mortilla, L. Amaducci, L. Bergamini, A. C. Bruni, J-F Foncin, F. Macciardi, M. P. Montesi, S. Sorbi, I. Rainero, L. Pinessi, R. J. Polinsky, D. Pollen, D. R. Crapper McLachlan, and P. H. St. George-Hyslop, "Analysis of the c-Fos Gene on chromosome 14 and the Promoter of the Amyloid Precursor Protein Gene in Familial Alzheimer's Disease" *Neurology* 43(1993): 2275–2279.

20. A. M. Saunders, W. J. Strittmatter, D. Schmechel, P. H. St. George-Hyslop, M. A. Pericak-Vance, S. H. Joo, B.L. Rosi, J. F. Gusella, D. R. Crapper-MacLachlan, M. J. Alberts, C. Hulette, B. Crain, D. Goldgaber, and A. D. Roses. "Association of Apolipoprotein E Allele E4 with Late-Onset Familial and Sporadic Alzheimer's Disease. *Neurology* 43 (1993): 1467–1472.

21. J. Poirier, J. Davignon, D. Bouthillier, S. Kogan, P. Bertrand, and S. Gauthier. "Apolipoprotein E Polymorphism and Alzheimer's Disease." *Lancet* 342 (1993): 697–699.

22. E. H. Corder, A. M. Saunders, W.J. Strittmatter, D. E. Schmechel, P. C. Gaskell, G. W. Small, A. D. Roses, J. L. Haines, and M. A. Pericak-

Vance. "Gene Dose of Apolipoprotein E Type 4 Allele and the Risk of Alzheimer's Disease in Late-Onset Families." *Science* 261 (1993): 921–923.

23. H. Payami, K. R. Montee, J. A. Kaye, T. D. Bird, C. -E. Yu, E. M. Wijsman, and G. D. Schellenberg. "Alzheimer's Disease, Apolipoprotein E4 and Gender." *Journal American Medical Association* 271 (1994): 1316–1317.

24. E. H. Corder, A. M. Saunders, N.J. Risch, W. J. Strittmatter, D. Schmechel, P. Gaskell, J. B. Rimmler, P. A. Locke, P. M. Conneally, K. E. Schmader, G. W. Small, A. D. Roses, J. L. Haines, and M. A. Pericak-Vance. "Apolipoprotein E Type 2 Allele Decreases the Risk of Late-Onset Alzheimer's Disease." *Nature Genetics* 7 (1994): 180–184.

25. M.-C. Chartier-Harlin, M. Parfitt, S. Legrain, J. Perez-Tur, T. Brousseau, A. Evans, C. Berr, O. Vidal, P. Roques, V. Gourlet, J.-C. Fruchart, A. Delacourte, M. Rossor, and P. Amouyel. "Apolipoprotein E, E4 Allele as a Major Risk Factor for Sporadic Early and Late-Onset Forms of Alzheimer's Disease: Analysis of the 19q13.2 Chromosomal Region." *Human Molecular Genetics* 3 (1994): 569–574.

26. Y. Namba, M. Tomonaga, H. Kawasaki, E. Otomo, and K. Ikeda. "Apolipoprotein E Immunoreactivity in Cerebral Amyloid Deposits and Neurofibrillary Tangles in Alzheimer's Disease and Kuru Plaque Amyloid in Creutzfeldt-Jakob Disease." *Brain Research* 541 (1991): 163–166.

27. T. Wisniewski, and B. Frangione. "Apolipoprotein E: A Pathological Chaperone Protein in Patients with Cerebral and Systemic Amyloid." *Neuroscience Letter* 135 (1992): 235–238.

28. W. J. Strittmatter, K. H. Weisgraber, D. Y. Huang, L.-M. Dong, G. S. Salvesen, M. Pericak-Vance, D. Schmechel, A. M. Saunders, D. Goldgaber, and A. D. Roses. "Binding of Human Apolipoprotein E to Synthetic Amyloid β Peptide: Isoform-Specific Effects and Implications for Late-Onset Alzheimer Disease." *Proceedings of the National Academy of Science* 90 (1993): 8098–8102.

29. T. Iwatsubo, A. Odaka, N. Suzuki, H. Mizusawa, N. Nukina, and Y. Ihara. "Visualization of Aβ42(43) and Aβ40 in Senile Plaques with End-Specific Aβ Monoclonals: Evidence that an Initially Deposited Species is Aβ42(43)." *Neuron* 13 (1994): 45–53.

30. D. A. Kirschner, H. Inouye, L. K. Duffy, A. Sinclair, M. Lind, and D. J. Selkoe. Synthetic Peptide Homologous to β Protein from Alzheimer Disease Forms Amyloid-Like Fibrils *in Vitro.*" *Proceedings of the National Academy of Science* 84 (1987): 6953–6957.

31. J. Ma, A. Yee, H. B. Brewer, Jr., S. Das, and H. Potter. "Amyloid-Associated Proteins α_1-antichymotrypsin and Apolipoprotein E Promote Assembly of Alzheimer β-Protein into Filaments." *Nature* 372 (1994): 92–94.

32. C. Behl, J. B. Davis, R. Lesley, and D. Schubert. "Hydrogen Peroxide Mediates Amyloid β Protein Toxicity." *Cell* 77 (1994): 817–827.

33. L. Meda, M. A. Cassatella, G. I. Szendrei, L. Otvos, Jr., P. Baron, M. Villalba, D. Ferrari, and F. Rossi. "Activation of Microglial Cells by β-amyloid Protein and Interferon-γ." *Nature* 374 (1995): 647–650.

34. A. D. Roses, "Apolipoprotein E Affects the Rate of Alzheimer Disease Expression: β-amyloid Burden Is a Secondary Consequence Dependent on APOE." *Journal of Neuropathology and Experimental Neurology* 53 (1994):429–437.

35. A. C. McKee, K. S. Kosik, and N. W. Kowall. "Neuritic Pathology and Dementia in Alzheimer's Disease." *Annals of Neurology* 30 (1991): 156–165.

36. S. E. Arnold, B. T. Hyman, J. Flory, A. R. Damasio, and G. W. Van Hoesen. "The Topographical and Neuroanatomical Distribution of Neurofibrillary Tangles and Neuritic Plaques in the Cerebral Cortex of Patients with Alzheimer's Disease." *Cerebral Cortex* 1 (1991): 103–116.

37. T. Wisniewski, M. Lalowski, A. Golabek, T. Vogel, and B. Frangione. "Is Alzheimer's Disease an Apolipoprotein E Amyloidosis?" *Lancet* 345 (1995): 956–958.

38. D. A. Drachman. "Is CNS Aging Inevitable? The New Frontier." Presidential Address. *American Neurological Association.* October 23, 1995.

Hannah's Gene

1. B.-S. Kerem, J. M. Rommens, J. A. Buchanan, D. Markiewicz, T. K. Cox, A. Chakravarti, M. Buchwald, and L.-C. Tsui. "Identification of the Cystic Fibrosis Gene: Genetic Analysis." *Science* 245 (1989): 1073–1080.

2. J. F. Gusella and Group. "A Novel Gene Containing a Trinucleotide Repeat That is Expanded and Unstable on Huntington's Disease Chromosomes." *Cell* 72 (1993): 971–983.

3. J. M. Rommens, B. Lin, G. B. Hutchinson, S. E. Andrew, Y. P. Goldberg, M. L. Glaves, R. Graham, V. Lai, J. McArthur, J. Nasir, J. Theilmann, H. McDonald, M. Kalchman, L. A. Clarke, K. Schappert, and M. R. Hayden. "A Transcription Map of the Region Containing the Huntington Disease Gene." *Human Molecular Genetics* 2 (1993): 901–907.

4. R. Sherrington, E. I. Rogaev, Y. Liang, E. A. Rogaeva, G. Levesque, M. Ikeda, H. Chi, C. Lin, G. Li, K. Holman, T. Tsuda, L. Mar, J.-F. Foncin, A. C. Bruni, M. P. Montesi, S. Sorbi, I. Rainero, L. Pinessi, L. Nee, I. Chumakov, D. Pollen, A. Brookes, P. Sanseau, R. J. Polinsky, W. Wasco, H.A.R. Da Silva, J. L. Haines, M. A. Pericak-Vance, R. E. Tanzi, A. D. Roses, P. E. Fraser, J. M. Rommens, and P. H. St. George-Hyslop. "Cloning of a Gene Bearing Missense Mutations in Early-Onset Familial Alzheimer's Disease." *Nature* 375 (1995): 754–760.

5. E. I. Rogaev, R. Sherrington, E. A. Rogaeva, G. Levesque, M. Ikeda, Y. Liang, H. Chi, C. Lin, K. Holman, T. Tsuda, L. Mar, S. Sorbi, B. Nacmias, S. Placentini, L. Amaducci, I. Chumakov, D. Cohen, L. Lannfelt, P. E. Fraser, J. M. Rommens, and P. H. St. George-Hyslop. "Familial Alzheimer's Disease in Kindreds with Missense Mutations in a Gene on Chromosome 1 Related to the Alzheimer's Disease Type 3 Gene." *Nature* 376 (1995): 775–778.

6. E. Levy-Lahad, E. M. Wijsman, E. Nemens, L. Anderson, K.A.B. Goddard, J. L. Weber, T. D. Bird, and G. D. Schellenberg. "A Familial Alzheimer's Disease Locus on Chromosome 1." *Science* 269 (1995): 970–973.

7. M. Barinaga. "Missing Alzheimer's Gene Found." *Science* 269 (1995): 917–918.

8. E. Levy-Lahad, W. Wasco, P. Poorkaj, D. M. Romano, J. Oshima, W. H. Pettingell, C. Yu, P. D. Jondro, S. D. Schmidt, K. Wang, A. C. Crowley, Y.-H. Fu, S. Y. Guenette, D. Galas, E. Nemens, E. M. Wijsman, T. D. Bird, G. D. Schellenberg, and R. E. Tanzi. "Candidate Gene for the Chromosome 1 Familial Alzheimer's Disease Locus." *Science* 269 (1995): 973–977.

9. C. F. Lippa, A. M. Saunders, T. W. Smith, J. M. Swearer, D. A. Drachman, B. Ghetti, L. Nee, D. Pulaski-Salo, D. Dickson, Y. Robitaille, C. Bergeron, B. Crain, M. D. Benson, M. Farlow, B. T. Hyman, P. St. George-Hyslop, A. D. Roses, and D. A. Pollen. "Familial and Sporadic Alzheimer's Disease: Neuropathology Cannot Exclude a Final Common Pathway." *Neurology* (1996): 406–412.

10. N. Suzuki, T. T. Cheung, X.-D. Cai, A. Odaka, L. Otvos, Jr., C. Eckman, T. E. Golde, and S. G. Younkin. "An Increased Percentage of Long Amyloid β Protein Secreted by Familial Amyloid β Protein Precursor (βAPP717) Mutants." *Science* 264 (1994): 1336–1339.

11. D. Games, D. Adams, R. Alessandrini, R. Barbour, P. Berthelette, C. Blackwell, T. Carr, J. Clemens, T. Donaldson, F. Gillespie, T. Guido, S. Hagopian, K. Johnson-Wood, K. Khan, M. Lee, P. Leibowitz, I. Lieberburg, S. Little, E. Masliah, L. McConlogue, M. Montoya-Zavala, L. Mucke, L. Paganini, E. Penniman, M. Power, D. Schenk, P. Seubert, B. Snyder, F. Soriano, H. Tan, J. Vitale, S. Wadsworth, B. Wolozin, and J. Zhao. "Alzheimer-Type Neuropathology in Transgenic Mice Overexpressing V717F β-amyloid Precursor Protein." *Nature* 373 (1995): 523–527.

12. D. Scheuner, T. Bird, M. Citron, L. Lannfelt, G. Schellenberg, D. Selkoe, M. Viitanen, and S. G. Younkin. "Fibroblasts from Carriers of Familial AD Linked to Chromosome 14 Show Increased Aβ Production." *Soc. Neuroscience Abstract* 21 (1995): 1500.

13. A. Tamaoka, P. E. Fraser, K. Ishii, N. Sahara, J. Rommens, M. Ikeda, R. Sherrington, G. Levesque, E. Rogaev, S. Shoji, L. Nee, D. Pollen, L. Farrer, P. H. St. George-Hyslop, and II, Mori. Increased long isoforms of amyloid-B-protein in brain of subjects with *presenilin* 1 mutations and Alzheimer's disease. Publication pending.

References

Abraham, C. R., Selkoe, D. J., and Potter, H., "Immunochemical Identification of the Serine Protease Inhibitor α_1-Antichymotrypsin in the Brain Amyloid Deposits of Alzheimer's Disease," *Cell* 52 (1988): 487–501.

Allen, G. E., *Thomas Hunt Morgan: The Man and His Science*. Princeton, N. J.: Princeton University Press, 1978.

Allsop, D., Landon, N, and Kidd, M. "The Isolation and Amino Acid Composition of Senile Plaque Core Protein." *Brain Research* 259 (1983): 348–352.

Alzheimer's Disease Research Group, J. Hardy et al. "Molecular Classification of Alzheimer's Disease." *Lancet* 337i (1991): 1342–1343.

Arber, W. "Promotion and Limitation of Genetic Exchange." *Science* 205 (1979): 361–365.

Barnes, D. M. "Defect in Alzheimer's Is On Chromosome 21." *Science* 235 (1987): 846–847.

Bick, K., Amaducci, L., and Pepeu, G., eds. *The Early Story of Alzheimer's Disease*. Padua, Italy: Lavonia Press; New York: Raven Press, 1987.

Bird, T. D., Sumi, S. M., Nemens, E. J., Nochlin, D., Schellenberg, G.,

Lampe, T. H., Sadovnick, A., Chui, H., Miner, G. W., and Tinklenberg, J. "Phenotypic Heterogeneity in Familial Alzheimer's Disease: A Study of 24 Kindreds." *Annals of Neurology* 25 (1989): 12–25.

Bird, T. D., Lampe, T. H., Nemens, E. K., Miner, G. W., Sumi, S. M., and Schellenberg, G. D. "Familial Alzheimer's Disease in American Descendants of the Volga Germans: Probable Genetic Founder Effect." *Annals of Neurology* 23 (1988): 25–31.

Bishop, Jerry E. and Waldholz, Michael. *Genome.* New York: Simon and Schuster, 1990.

Bodmer, W. F. "The William Allan Memorial Award Address: Gene Clusters, Genome Organization, and Complex Phenotypes: When the Sequence Is Known, What Will It Mean?" *American Journal of Human Genetics* 33 (1981): 664–682.

Botstein, D., White, R. L., Skolnick, M., and Davis, R. W. "Construction of a Genetic Linkage Map in Man Using Restriction Fragment Length Polymorphisms." *American Journal of Human Genetics* 32 (1980): 314–331.

Breitner, J. C. S., Folstein, M. F., and Murphy, E. A. "Familial Aggregation in Alzheimer Dementia. I. A Model for Age-dependent Expression of an Autosomal Dominant Gene." *Journal of Psychiatric Research* 20 (1986): 31–43.

Brown, P., Goldfarb, L. G., and Gajdusek, D. C. "The New Biology of Spongiform Encephalopathy: Infectious Amyloidosis with a Genetic Twist." *Lancet* 337 (1991): 1019–1022.

Brown, P. "Central Nervous System Amyloidoses: A Comparison of Alzheimer's Disease and Creutzfeldt-Jakob Disease." *Neurology* 39 (1989): 1103–1105.

Brown, P., Coker-Vann, M., Pomeroy, K., Franko, M., Asher, D. M., Gibbs, C. J., and Gajdusek, D. C. "Diagnosis of Creutzfeldt-Jacob Disease by Western Blot Identification of Marker Protein in Human Brain Tissue." *New England Journal of Medicine* 314 (1986): 547–551.

Bueler, H., Fischer, M., Lang, Y., Bluethmann, H., Lipp, H.-P., DeArmond, S. J., Prusiner, S. B., Aguet, M., and Weissmann, C. "Normal Development and Behaviour of Mice Lacking the Neuronal Cell-Surface PrP Protein." *Nature* 356 (1992): 577–598.

Burger, P. C., and Vogel, S. F. "The Development of the Pathologic Changes of Alzheimer's Disease and Senile Dementia in Patients with Down's Syndrome." *American Journal of Pathology* 73 (1973): 457–476.

Burghes, A. H. M., Logan, C., Hu, X., Belfall, B., Warton, R. G., and Ray, P. N. "A cNDA Clone from the Duchenne/Becker Muscular Dystrophy Gene." *Nature* 328 (1987): 434–437.

Bush, A. I., Martins, R. N., Rumble, B., Moir, R., Fuller, S., Milward, E., Currie, J., Ames, D., Weidemann, A., Fischer, P., Multhaup, G., Beyreuther, K., and Masters, C. L. "The Amyloid Precursor Protein of Alzheimer's Disease Is Released by Human Platelets." *Journal of Biological Chemistry* 265 (1990): 15977–15983.

Cameron, J. R., Loh, E. Y., and Davis, R. W. "Evidence for Transposition of

Dispersed Repetitive DNA Families in Yeast." *Cell* 16 (1979): 739–751.

Chartier-Harlin, M.-C., Crawford, F., Houlden, H., Warren, A., Hughes, D., Fidani, L., Goate, A., Rossor, M., Roques, P., Hardy, J., and Mullan. M. "Earlier-Onset Alzheimer's Disease Caused by Mutations at Codon 717 of the β-amyloid Precursor Protein Gene." *Nature* 353 (1991): 844–846.

Cherfas, J. "Mad Cow Disease: Uncertainty Rules." *Science* 249 (1990): 1492–1493.

Chesebro, B. "PrP and the Scrapie Agent." *Nature* 356 (1992): 560.

Coleman, P., ed. September issue of *Neurobiology of Aging* 13 (1992): No. 5.

Conneally, P. M. "A First Step Toward a Molecular Genetic Analysis of Amyotrophic Lateral Sclerosis." *New England Journal of Medicine* 324 (1991): 1430–1432.

Constantinidis, J., Garrone, G., and de Ajuriaguerra, J. "L'heredite des demences de l'age avance." *Encephale* 51 (1962): 301–344.

Cook, R., Ward, B. E., and Austin, J. H. "Studies in Aging of the Brain. IV. Familial Alzheimer's Disease: Relation to Transmissible Dementia, Aneuploidy and Microtubular Defects." *Neurology* 29 (1979): 1402–1412.

Cook-Deegan, R. M. *The Gene Wars: Science, Politics and the Human Genome.* New York: Norton, 1993.

Crick, F. H. C., Barnett, L., Brenner, S. and Watts-Tobin, R. J. "General Nature of the Genetic Code for Proteins." *Nature* 192 (1961): 1227–1232.

Custer, B. G. *I Chose to Continue.* Minneapolis: Lakeland Color Press, 1991.

Davies, K. E., Pearson, P. L., Harper, P. S., Murray, J. M., O'Brien, T., Sarfarazi, M., and Williamson, R. "Linkage Analysis of Two Cloned DNA Sequences Flanking the Duchenne Muscular Dystrophy Locus on the Short Arm of the Human X Chromosome." *Nucleic Acids Research* 11 (1983): 2303–2312.

Delabar, J. M., Goldgaber, D., Lamour, Y., Nicole, A., Huret, J.-L., de Grouchy, J., Brown, P., Gajdusek, E. C., and Sinet, P.-M. "β Amyloid Gene Duplication in Alzheimer's Disease and Karyotypically Normal Down Syndrome." *Science* 235 (1987): 1390–1392.

Dubnow, S. M., and Friedlander, I. *The History of the Jews in Russia and Poland*, Vols. I, II. Philadelphia: Jewish Publication Society of America, 1918.

Ellis, W. G., McCullick, J. R., and Corley, C. L. "Presenile Dementia in Down's Syndrome: Ultrastructural Identity with Alzheimer's Disease." *Neurology* 24 (1974): 101–106.

Esch, F. S., Kleim, P. S., Beattie, E. C., Blacher, R. W., Culwell, A. R., Oltersdorf, T., McClure, D., and Ward, P. J. "Cleavage of Amyloid β-Peptide During Constitutive Processing of Its Precursor." *Science* 248 (1990): 1122–1124.

Evans, D. A., Funkenstein, H., Albert, M. S., Scherr, P. A., Cook, N. R., Chown, M. J., Hebert, L. E., Henneskens, C. H., and Taylor, J. O. "Prevalence of Alzheimer's Disease in a Community Population of

Older Persons." *Journal of the American Medical Association* 262 (1989): 2551–2556.

Feldman, R. G., Chandler, K. A., Levy, L. L., and Glaser, G. H. "Familial Alzheimer's Disease." *Neurology* 13 (1963): 811–824.

Fiers, W., Contreras, R., Haegeman, G., Rogiers, R., Van de Voorde, A., Van Heuverswyn, H., Van Herreweghe, J., Volckgert, G., and Ysebaert, M. "Complete Nucleotide Sequence of SV40 DN." *Nature* (London) 273 (1978): 113–120.

Fisher, R., and Bennett, J. H., eds. *Experiments in Plant Hybridization: Mendel's Original Papers in English Translation with Commentary and Assessment.* London: Oliver and Boyd, 1965.

Foncin, J.-F., Salmon, D., and Bruni, A. C. "Extended Kindreds as a Model for Research on Alzheimer's Disease." In *Genetics and Alzheimer's disease,* edited by P. M. Sinet, Y. Lamour, and Y. Christen. New York: Springer-Verlag: 1988.

Foncin, J.-F., Salmon, D., Supino-Viterbo, V., Feldman, R. G., Macchi, G., Mariotti, P., Scoppetta, C., Caruso, G., and Bruni, A. C. "Demence Presenile d'Alzheimer transmise dans une famille etendue." *Revue Neurologique* (Paris) 141 (1985): 194–202.

Fox, P. "From Senility to Alzheimer's Disease: The Rise of the Alzheimer's Disease Movement." *The Milbank Quarterly* 67 No. 1 (1989): 58–102.

Frommelt, P., Schnabel, R., Kuhne, W., Nee, L. E., and Polinsky, R. J. "Special Communication: Familial Alzheimer Disease: A Large, Multi-generation German Kindred." *Alzheimer Disease and Associated Disorders* 5 (1991): 36–43.

Gajdusek, D. C., "Unconventional Viruses and the Origin and Disappearance of Kuru." *Science* 197 (1977): 943–960.

Gajdusek, D. C., and Zigas, V. "Kuru." *American Journal of Medicine* 26 (1959): 442–469.

Gajdusek, D. C., and Zigas, V. "Degenerative Disease of the Central Nervous System in New Guinea: The Endemic Occurrence of "Kuru" in the Native Population." *New England Journal of Medicine* 257 (1957): 974–978.

Gardella, J. E., Ghiso, J., Gorgone, G. A., Marratta, D., Kaplan, A. P., Frangione, B., and Gorevic, P. D. "Intact Alzheimer Amyloid Precursor Protein (APP) Is Present in Platelet Membranes and Is Encoded by Platelet mRNA." *Biochemical and Biophysical Research Communications* 173 (1990): 1292–1298.

Gaupp, R. "Alois Alzheimer." *Muenchener Medizinische Wochenschrift,* February 1916, pp.195–196.

Gibbs, C. J., Jr., and Gajdusek, D. C. "Infection as the Etiology of Spongiform Encephalopathy (Creutzfeldt-Jakob Disease)." *Science* 165 (1969): 1023–1025.

Glenner, G. G., and Wong, C. W. "Alzheimer's Disease and Down's Syndrome: Sharing of a Unique Cerebrovascular Amyloid Fibril Protein." *Biochemical and Biophysical Research Communications* 122 (1984): 1131–1135.

Glenner, G. G., and Wong, C. W. "Alzheimer's Disease: Initial Report of the Purification and Characterization of a Novel Cerebrovascular

Amyloid Protein." *Biochemical and Biophysical Research Communications* 120 (1984): 885–890.

Goate, A., Charter-Harlin, M. C., Mullan, M., Brown, J., Crawford, F., Fidani, L., Giuffra, L., Haynes, A., Irving, N., James, L., Mant, R., Newton, P., Rooke, K., Roques, P., Talbot, C., Pericak-Vance, M., Roses, A., Williamson, R., Rossor, M., Owen, M., and Hardy, J. "Segregation of a Mis-sense Mutation in the Amyloid Precursor Protein Gene with Familial Alzheimer's Disease." *Nature* 349 (1991): 704–706.

Goldgaber, D., Lerman, M. I., McBride, O. W., Saffiotti, U., and Gajdusek, D. C. "Characterization and Chromosomal Localization of a cDNA Encoding Brain Amyloid of Alzheimer's Disease." *Science* 235 (1987): 877–880.

Goodman, H., Olson, M., and Hall, B. "Nucleotide Sequence of a Mutant Eukaryotic Gene: The Yeast Tyrosine Inserting Ochre Suppressor SUP4-0." *Proceedings of National Academy of Science* 74 (1977): 5453–5457.

Gorevic, P. D., Gardella, J. E., Newman, P. J., Frangione, B. "The Platelets: A Potential Source of Amyloid Protein in Alzheimer's Disease." *Clinical Research* 38 (1991): 200A.

Goudsmit, J., White, B. J., Whitekamp, L. R., Keats, B. J. B., Murrow, C. H., and Gadjusek, D. C. "Familial Alzheimer's Disease in Two Kindreds of the Same Geographic and Ethnic Origin." *Journal of Neurological Sciences* 49 (1981): 79–89.

Goudsmit, J., Morrow, C. H., Asher, D. M., Yanagihara, R. T., Masters, C. L., Gibbs, C. J., Jr., and Gajdusek, D. C., "Evidence for and against the Transmissibility of Alzheimer's Disease." *Neurology* 30 (1980): 945–950.

Grodzicker, T., Williams, J., Sharp, P., and Sambrook, J. "Physical Mapping of Temperature Sensitive Mutations of Adenoviruses." Cold Spring Harbor Symposium. *Quantitative Biology* 39 (1974): 439–446.

Gusella, J. F., Keys, C., Varsanyi-Breiner, A., Kao F.-T., Jones, C., Puck, T. T., and Housman, D. *Proceedings of the National Academy of Science* 77 (1980): 2829–2833.

Gusella, J. F., Wexler, N. S., Conneally, P. M., Naylor, S. L., Anderson, M. A., Tanzi, R. E., Watkins, P. C., Ottina, K., Wallace, M. R., Sakagucki, A. Y., Young, A. B., Shoulson, I., Bonilla, E., and Martin, J. B. "A Polymorphic DNA Marker Genetically Linked to Huntington's Disease." *Nature* 306 (1983): 234–238.

Haan, J., Lanser, J. B. K., Zijderveld, I., van der Sols, I. G. R., and Roos, R. A. C. "Dementia in Hereditary Cerebral Hemorrhage with Amyloidosis-Dutch Type." *Archives of Neurology* 47 (1990): 965–968.

Hadlow, W. J. "Scrapie and Kuru." *Lancet* 2 (1959): 289–290.

Haldane, J.B. S., and Smith, C. A. B. "A New Estimate of the Linkage Between the Genes for Colour-Blindness and Haemophilia in Man." *Annals of Eugenics* 14 (1947): 1031.

Hamilton, W. D. "Selection of Selfish and Altruistic Behavior in Some Extreme Models." In *Man and Beast: Comparative Social Behavior*, edited by J. F. Eisenberg and W. S. Dillon eds., pp. 57–91.

Hamilton, W. D. "The Genetical Theory of Social Behavior. I, II." *Journal of Theoretical Biology* 7 (1964): 1–52, also Washington, D.C.: Smithsonian Institution Press, 1971.

Hamos, J. E., Oblas, B., Pulaski-Salo, D., Welch, W. J., Bole, D. G., and Drachman, D. A., "Expression of Heat Shock Proteins in Alzheimer's Disease," *Neurology* 41 (1991): 345–350.

Hardy, J. A., and Higgins, G. A. "The Amyloid Cascade Hypothesis." *Science* 256 (1992): 184–185.

Hardy, J., and Allsop, D. "Amyloid Deposition as the Central Event in the Aetiology of Alzheimer's Disease." *Trends in Pharmacological Science* 12 (1991): 383–388.

Hendriks, L., van Duijn, C. M., Cras, P., Cruts, M., Van Hul, W., van Harskamp, F., Warren, A., McInnis, M. G., Antonarakis, S. E., Martin, J.-J., Hofman, A., and Van Broeckhoven, C. "Presenile Dementia and Cerebral Haemorrhage Linked to a Mutation at Codon 692 of the β-Amyloid Precursor Protein Gene." *Nature Genetics* 1 (1992): 218–221.

Heston, L. L., Lowther, D. L., and Leventhal, C. M. "Alzheimer's Disease: A Family Study." *Archives of Neurology* 15 (1966): 225–233.

Heston, L. L., and Mastri, A. R. "The Genetics of Alzheimer's Disease: Associations with Hematologic Malignancy and Down's Syndrome." *Archives of General Psychiatry* 34 (1977): 976–981.

Heston, L. L. "Alzheimer's Disease, Trisomy 21, and Myeloproliferative Disorders: Associations Suggesting a Genetic Diathesis." *Science* 196 (1977): 322–323.

Heyman, A., Wilkinson, W. E., Hurwitz, B. U., Schmechel, D., Sigmon, A. H., Weinberg, T., Helms, M. F., and Swift, M. "Alzheimer's Disease: Genetic Aspects and Associated Clinical Disorders." *Annals of Neurology* 14 (1983): 507–515.

Hoff, P., and Hippius, H. "Alzheimer Alois, 1864–1915. Ein Uberblick uber Leben und Werk anlßslich seines 125." *Geburstages, Nervenarzt* 60 (1989):332–337.

Holden, C. "Politics and Soviet Psychiatry." *Science* 239 (1988): 551–553.

Hollmann, M., Hartley, M., and Heinemann, S. "Ca^{2+} Permeability of KA-AMPA-gated Glutamate Receptor Channels Depends on Subunit Composition." *Science* 252 (1991): 851–853.

Hooper, C., "Of Rumors, Reviews, and Rip-Offs in Alzheimer's Research," *The Journal of National Institutes of Health Research* 4 (1992): 34.

Housman, D., and Gusella, J. "Molecular Genetic Approaches to Neural Degenerative Disorders." In *Molecular Genetic Neuroscience*, edited by F. O. Schmitt, S. J. Bird, and F. E. Bloom. New York: Raven Press, 1982, pp. 415–422.

Hsiao, K., Baker, H. F., Crow, T. J., Poulter, M., Owen, F., Terwilliger, J. D., Westaway, D., Ott, J., and Prusiner, S. B. "Linkage of a Prion Protein Mis-sense Variant to Gerstmann-Strausler Syndrome." *Nature* 338 (1989): 342–345.

Hutchinson, C., III, Newbold, J., Potter, S., and Edgell, M. "Maternal Inheritance of Mammalian Mitochondrial DNA." *Nature* 251 (1974): 536.

Jeffreys, A. J. "DNA Sequence Variants in the $^{G}\gamma$-, $^{A}\gamma$-, δ- and β-globin Genes of Man." *Cell* 18 (1979): 1–10.

Jervis, G. A. "Early Senile Dementia in Mongoloid Idiocy." *American Journal of Psychiatry* 105 (1948): 102–106.

Joachim, C. L., Mori, H., and Selkoe, D. J. "Amyloid β-protein Deposition in Tissues Other Than Brain in Alzheimer's Disease." *Nature* 341 (1989): 226–230.

Jordan, E. "Invited Editorial: The Human Genome Project: Where Did It Come From, Where Is It Going?" *American Journal of Human Genetics* 51 (1992): 1–6.

Judson, H. F. *The Eighth Day of Creation.* New York: Simon and Schuster, 1979.

Kallmann, F. J. In *Mental Disorders in Later Life,* 2nd ed. Revised by O. J. Kaplan. Stanford, Calif.: Stanford University Press, 1956.

Kan, Y. W., and Dozy, A. M. "Polymorphisms of DNA Sequence Adjacent to Human β-globin Structural Gene: Relationship to Sickle Mutation." *Proceedings of the National Academy of Science* 75 (1978): 5631–5635.

Kan, Y., and Dozy, A. "Antenatal Diagnosis of Sickle-cell Anaemia by DNA Analysis of Amniotic-fluid Cells." *Lancet* 2 (1978): 910–912.

Kang, J., Lemaire, H.-G., Unterbeck, A., Salbaum, J. M., Masters, C. L., Grzeschik, K.-H., Multhaup, G., Beyreuther, K., and Müller-Hill, B. "The Precursor of Alzheimer's Disease Amyloid A4 Protein Resembles a Cell-Surface Receptor." *Nature* 325 (1987): 733–736.

Katzman, R., Terry, R. D., and Bick, K. L. "Recommendations of the Nosology, Epidemiology, and Etiology and Pathophysiology Commissions of the Workshop—Conference on Alzheimer's Disease, Senile Dementia and Related Disorders." In *Alzheimer's Disease: Senile Dementia and Related Disorders* (*Aging,* Vol. 7), edited by R. Katzman, R. D. Terry, and K. L. Bick. New York: Raven Press, 1978, pp. 579–585.

Katzman, R., and Karasu, T. "Differential Diagnosis in Dementia." In *Neurological and Sensory Disorders in the Elderly,* edited by W. Fields. New York: Stratton Intercontinental Medical Book Corp., 1975, pp. 103–134.

Katzman, R., Suzuki, K., and Korey, S. "Chemical Studies on Alzheimer's Diease." *Neuropathology and Experimental Neurology* 24 (1965): 211–224.

Katzman, R. "The Prevalence and Malignancy of Alzheimer's Disease." *Archives of Neurology* 33 (1976): 217–218.

Kawabata, S., Higgins, G. A., and Gordon, J. W. "Amyloid Plaques, Neurofibrillary Tangles and Neuronal Loss in Brains of Transgenic Mice Overexpressing a C-terminal Fragment of Human Amyloid Precursor Protein." *Nature* 354 (1991): 476–478. See also retraction of this paper in *Nature* 356 (1992): 23, and in *Nature* 356 (1992): 265.

Kidd. M. "Paired Helical Filaments in Electron Microscopy of Alzheimer's Disease." *Nature* 197 (1963): 192–193.

King, M.-C., and Wilson, A. C. "Evolution at Two Levels in Humans and Chimpanzees." *Science* 188 (1975): 107–116.

Kingston, H. M., Harper, P. S., Pearson, P. L., Davies, K. E., Williamson, R., and Page, D. "Localisation of Gene for Becker Muscular Dystrophy." *Lancet* 2 (1983): 1200.

Koch, F. C. *The Volga Germans*. University Park: Pennsylvania State University Press, 1977.

Koenig, M., Hoffman, E. D., Bertelson, C. J., Monaco, A. P., Feener, C., and Kunkel, L. M. "Complete Cloning of the Duchenne Muscular Dystrophy (DMD) cDNA and Preliminary Genomic Organization of the DMD Gene in Normal and Affected Individuals." *Cell* 50 (1987): 509–517.

Kowall, N. W., Beal, M. F., Busciglio, J., Duffy, L. K., and Yankner, B. A. "An *in vivo* Model for the Neurodegenerative Effects of β-Amyloid and Protection by Substance P." *Proceedings of the National Academy of Science* 88 (1991): 7247–7251.

Kozlowski, M. R., Spanoyannis, A., Manly, S. P., Fidel, S. A., and Neve, R. L. "The Neurotoxic Carboxy-Terminal Fragment of the Alzheimer's Amyloid Precursor Binds Specifically to a Neuronal Cell Surface Molecule: PH Dependence of the Neurotoxicity and the Binding." *Journal of Neuroscience* 12 (1992): 1679–1687.

Krigman, M. R., Feldman, R. G., and Bensch, K. "Alzheimer's Presenile Dementia: A Histochemical and Electron Microscopic Study." *Laboratory Investigation* 14 (1965): 381–396.

Kropotkin, P. "Mutual Aid: A Factor of Evolution." New York: McClure Phillips and Co., 1902.

Larsson, T, Sjogren, T., and Jacobson, G. "A Clinical, Sociomedical and Genetic Study." *Acta Psychiatrica et Neurologica Scandinavica Supplement* 39, suppl. 167 (1963).

Lauter, H., "Some Remarks on the Second Medical Faculty in Munich and on the Life and Scientific Achievements of Alois Alzheimer," Lecture given at a Psychiatric Symposium in Munich, October 4, 1990.

Lawn, R. M., Fritsch, E. F., Parker, R. C., Blake, G., and Maniatis, T. "The Isolation and Characterization of Linked δ- and β-globin Genes from a Cloned Library of Human DNA." *Cell* 15 (1978): 1157–1174.

Lawrence, J. B., "A Fluorescece In Situ Hybridization Approach for Gene Mapping and the Study of Nuclear Organization," *Genome Analysis Volume 1: Genetic and Physical Mapping* (1990): 1–38. Cold Spring Harbor Laboratory Press.

Leder, P. "Can the Human Genome Project Be Saved From Its Critics and Itself?" *Cell* 63 (1990): 1–3.

Lejeune, J., Turpin, B., and Gautier, M. "Chromosomic Diagnosis of Mongolism." *Archives Françaises de Pediatrie* 16 (1959): 962–963.

Levy, E., Carman, M. D., Fernandez-Madrid, I. J., Power, M. D., Lieberburg, I., Van Duinen, S. G., Bots, G. T. A. M., Luydendijk, W., and Frangione, B. "Mutation of the Alzheimer's Disease Amyloid Gene in Hereditary Cerebral Hemorrhage, Dutch Type." *Science* 248 (1990): 1124–1126.

Lewey, F. H. "Alois Alzheimer." In *The Founders in Neurology*, compiled and edited by Webb Haymaker and Frances Schiller, 2nd ed. Springfield, Ill.: Charles C Thomas, 1970, pp. 315–318.

Lewin, B., *Genes IV*, New York: Oxford University Press, 1991.

Maniatis, T., Hardison, R. C., Lacy, E., Lauer, J., O'Connell, C., and Quon, D. "The Isolation of Structural Genes from Libraries of Eucaryotic DNA." *Cell* 15 (1978): 687–701.

Maragos, W. F., Greenamyre, J. T., Penney, J. B., and Young, A. "Glutamate Dysfunction in Alzheimer's Disease: An Hypothesis." *TINS* 10 (1987): 65–68.

Marks, J. L. "The Cystic Fibrosis Gene Is Found." *Science* 245 (1989): 923–925.

Marshall, E. "Data Sharing: A Declining Ethic?" *Science* 248 (1990): 952–957.

Martin, J. B. "Huntington's Disease: Genetically Determined Cell Death in the Mature Human Nervous System." *Molecular Genetic Neuroscience*, edited by F. O. Schmitt, S. J. Bird, and F. E. Bloom. New York: Raven Press, 1982.

Martin, J-J., Gheuens, J., Bruyland, M., Cras, P., Vandenberghe, A., Masters, L., Beyreuther, K., Dom, R., Ceuterick, C., Lübke, U., Van Heuverswijn, H., DeWinter, G., and Van Broeckhoven, C., "Early-Onset Alzheimer's Disease in 2 Large Belgian Families," *Neurology* 41 (1991): 62–68.

Marx, J. L. "Multiplying Genes by Leaps and Bounds." *Science* 240 (1988): 1408–1410.

Marx, J. "Alzheimer's Debate Boils Over." *Science* 257 (1992): 1336–1338.

Marx, J. "Boring in on β-Amyloid's Role in Alzheimer's." *Science* 255 (1992): 688–689.

Masters, C. L., Multhaup, G., Simms, G., Pottgiesser, J., Martins, R. N., and Beyreuther, K. "Neuronal Origin of a Cerebral Amyloid: Neurofibrillary Tangles of Alzheimer's Disease Contain the Same Protein as the Amyloid of Plaque Cores and Blood Vessels." *European Molecular Biology Organization Journal* 4 (1985): 2757–2763.

Masters, C. L., Gajdusek, D. C., and Gibbs, C. J., Jr. "The Familial Occurrence of Creutzfeldt-Jakob Disease and Alzheimer's Disease." *Brain* 104 (1981): 535–559.

Masters, C. L., Simms, G., Weinman, N. A., Multhaup, G., McDonald, B. L., and Beyreuther, K. "Amyloid Plaque Core Protein in Alzheimer Disease and Down Syndrome." *Proceedings of the National Academy of Science* 82 (1985): 4245–4249.

Maxam, A., and Gilbert, W. "A New Method for Sequencing DNA." *Proceedings of the National Academy of Science* 74 (1977): 5463–5467.

Mayer-Gross, W. "Recent Progress in Psychiatry, Arteriosclerotic, Senile and Presenile Dementia." *Journal of Mental Science* 90 (1944): 316–327.

McEntel, W. J., and Crook, T. J. "Age-Associated Memory Impairment: A Role for Catecholamines." *Neurology* 40 (1990): 526–530.

McKusick, V. A. and Ruddle, F. H. "The Status of the Gene Map of the Human Chromosomes." *Science* 196 (1977): 390–405.

Merz, P., Somerville, R. A., Wisniewski, H. M., Manuelides, L., and Manuelides, E. E. "Scrapie-Associated Fibrils in Creutzfeldt-Jacob Disease." *Nature* 306 (1983): 474–478.

Merz, P. A., Rohwer, R. B., Kascsak, R., Wisniewski, H. M., Somerville, R. A., Gibbs, C. J., and Gajdusek, D. C. "Infection Specific Particle from the Unconventional Slow Virus Diseases." *Science* 225 (1984): 437–440.

Merz, P. A., Somerville, R. A., Wisniewski, A. M., and Iqbal, K. "Abnormal

Fibrils from Scrapie-Infected Brain." *Acta Neuropathologica Berlin* 54 (1981): 63–74.

Mohs, R. C., Breitner, J. C. S., Silverman, J. M., and Davis, K. L. "Alzheimer's Disease: Morbid Risk Among First-Degree Relatives Approximates 50% by 90 Years of Age." *Archives of General Psychiatry* 44 (1987): 405–408.

Monaco, A. P., Neve, R. L., Colletti-Feener, C., Bertelson, C. J., Kurnit, D. M., and Kunkel, L. M. "Isolation of Candidate cDNAs for Portions of the Duchenne Muscular Dystrophy Gene." *Nature* 323 (1986): 646–650.

Morton, N. E. "Further Scoring Types in Sequential Linkage Tests, with a Critical Review of Autosomal and Partial Sex Linkage in Man." *American Journal of Human Genetics* 9 (1957): 55–75.

Mullan, M., Crawford, F., Axelman, K., Houlden, H., Lilius, L., Winblad, B., and Lannfelt, L. "A Pathogenic Mutation for Probable Alzheimer's Disease in the APP Gene at the N-terminus of β-Amyloid." *Nature Genetics* 1 (1992): 345–347.

Mullis, K. B. "The Unusual Origin of the Polymerase Chain Reaction." *Scientific American* 262 (1990): 56–65.

Murray, J. M., Davies, K. E., Harper, P. S., Meredith, L., Mueller, C. R., and Williamson, R. "Linkage Relationship of a Cloned DNA Sequence on the Short Arm of the X Chromosome to Duchenne Muscular Dystrophy." *Nature* 300 (1982): 69–71.

Murrell, J., Farlow, M., Ghetti, B., and Benson, M. D. "A Mutation in the Amyloid Precursor Protein Associated with Hereditary Alzheimer's Disease." *Science* 254 (1992): 97–99.

Naruse, S., Igaraski, S., Aoki, K., Kaneko, K., Iihara, M., Miyatake, T., Kobayashi, H., Inuzuka, T., Shimizu, T., Kojima, T., and Tsuji, S. "Mis-sense Mutation Val → Ile in Exon 17 of Amyloid Precursor Protein Gene in Japanese Familial Alzheimer's Disease." *Lancet* 337 (1991):978.

Nathans, D. "Restriction Endonucleases, Simian Virus 40, and the New Genetics." *Science* 206 (1979): 903–909.

Nee, L. E., Polinksy, R. J., Eldridge, R., Weingartner, H., Smallberg, S., and Ebert, M. "A Family with Histologically Confirmed Alzheimer's Disease." *Archives of Neurology* 40 (1983): 203–209.

Newton, R. D. "The Identity of Alzheimer's Disease and Senile Dementia and Their Relationship to Senility." *Journal of Mental Science* 94 (1948): 225–249. (Please refer to Newton's bibliography for references to the early work cited in his paper.)

Olson, M. I., and Shaw, C. M. "Presenile Dementia and Alzheimer's Disease in Mongolism." *Brain* 92 (1969): 147–156.

Orel, V., *Gregor Mendel*, New York: Oxford University Press, 1984.

Pauling, L., and Delbrück, M. "The Nature of the Intermolecular Forces Operative in Biological Processes." *Science* 92 (1940): 77–79.

Pericak-Vance, M. A., Conneally, P. M., Merritt, A. D., Roos, R. P., Vance, J. M., Yu, P. L., Norton, J. A., Jr., and Antel, J. P. "Genetic Linkage in Huntington's Disease." *Advances in Neurology* 23 (1979): 59–72.

Pericak-Vance, M. A., Bebout, J. L., Gaskell, P. C., Jr., Yamaoka, L. H., Hung, W. Y., Alberts, M. J., Walker, A. P., Bartlett, R. J., Haynes, C.

A., Welsh, K. A., Earl, N, L., Heyman, A., Clark, C. M., and Roses, A. D. "Linkage Studies in Familial Alzheimer's Disease: Evidence for Chromosome 19 Linkage." *American Journal of Human Genetics* 48 (1991): 1034–1050.

Petes, T. K., and Botstein, D. "Simple Mendelian Inheritance of the Reiterated Ribosomal DNA of Yeast." *Proceedings of the National Academy of Science* 74 (1977): 5091–5095.

Podlisny, M. B., Lee, G., and Selkoe, J. "Gene Dosage of the Amyloid β-Precursor Protein in Alzheimer's Disease." *Science* 238 (1987): 669–671.

Potter, S., Newbold, J., Hutchinson, C., III, and Edgell, M. "Specific Cleavage Analysis of Mammalian Mitochondrial DNA." *Proceedings of the National Academy of Science* 72 (1975): 4496–4500.

Pratt, R. T. C. "The Genetics of Alzheimer's Disease." In *Alzheimer's Disease and Related Conditions*: A Ciba Foundation Symposium, edited by G. E. W. Wolstenholme and Maeve O'Connor. London: Churchill, 1970.

Prusiner, S. B. "Molecular Biology of Prion Proteins." *Science* 252 (1991): 1515–1522.

Prusiner, S. B. "Prions and Neurodegenerative Diseases." *New England Journal of Medicine* 317 (1987): 1571–1581.

Race, R. R., and Sanger, R. *Blood Groups in Man*, 5th ed. Philadelphia: F. A. Davis Co., 1968.

Reddy, V. B., Thimmappaya, B., Dhar, R., Subramanian, N., Zain, B. S., Pan, J., Ghosh, P. K., Celma, M.L., and Weissman, S. M. "The Genome of Simian Virus 40." *Science* 200 (1978): 494–502.

Rennie, J., "The Mice That Missed." *Scientific American* 266 (1992): 20–21.

Renvoize, E. B., Hambling, M. H., Pepper, M. D., and Rajah, S. M. "Possible Association of Alzheimer's Disease with HLA-BW15 and Cytomegalous Infection." *Lancet* 1 (1979): 1238.

Risch, N., "Genetic Linkage: Interpreting Lod Scores." *Science* 255 (1992): 803–804.

Robakis, N. K., Wisniewski, H. M., Jenkins, E. C., Devine-Gage, E. A., Houck, G. E., Yao, X.-L., Ramakrishna, N., Wolfe, G., Silverman, W. P., and Brown, W. T. "Chromosome 21q21 Sublocalisation of Gene Encoding Beta-Amyloid Peptide in Cerebral Vessels and Neuritic (Senile) Plaques of People with Alzheimer Disease and Down Syndrome." *Lancet* (February 14, 1987): 384–385.

Roberts, L. "Huntington's Gene: So Near, Yet So Far." *Science* 247 (1990): 624–627.

Roberts, L. "Genome Backlash Going Full Force." *Science* 248 (1990): 804.

Roberts, L. "Race for Cystic Fibrosis Gene Nears End." *Science* 240 (1988): 282–285.

Roberts, L. "Genome Project: An Experiment in Sharing." *Science* 248 (1990): 953.

Roberts, L. "The Race for the Cystic Fibrosis Gene." *Science* 240 (1988): 141–144.

Roberts, L. "The Rush to Publish." *Science* 251 (1991): 260–263.

Rogaev, E. I., Lukiw, W. J., Vaula, G., Haines, J. L., Rogaeva, E. A., Tsuda, T., Alexandrova, N., Liang, Y., Mortilla, M., Amaducci, L., Bergamini, L., Bruni, A. C., Foncin, J-F., Macciardi, F., Montesi, M. P., Sorbi, S.,

Rainero, I., Pinessi, L., Polinsky, R. J., Pollen, D., Crapper McLachlan, D. R., and St. George-Hyslop, P. H., "Analysis of the c-FOS Gene on Chromosome 14 and the Promoter of the Amyloid Precursor Protein Gene in Familial Alzheimer's Disease." *Neurology* 43(1993): 2275–2279.

Roses, A. D., Pericak-Vance, M. A., Haynes, C. S., Haines, J. L., Gaskell, P. A., Yamaoka, L. H., Hung, W.-Y., Clark, C. M., Alberts, M. J., Lee, J. E., Siddique, T., and Heyman, A. L. "Genetic Linkage Studies in Alzheimer's Disease (AD)." *Neurology* 39, suppl. 1 (1988): 173; *Neurology* 38, suppl. 1 (1988): 173.

Roses, A. D., and Conneally, P. M. "Alzheimer's Disease Cell Bank." *Science* 252 (1991): 14.

Sanger, F., and Coulson, A. R. "A Rapid Method for Determining Sequences in DNA by Primed Synthesis with DNA Polymerase." *Journal of Molecular Biology* 94 (1975): 441–448.

Schellenberg, G. D., Bird, T. D., Wijsman, E. M., Moore, D. K., Boehnke, M., Bryant, E. M., Lampe, T. H., Nochlin, D., Sumi, S. M., Deeb, S. S., Beyreuther, K., and Martin, G. M. "Absence of Linkage of Chromosome 21q21 Markers to Familial Alzheimer's Disease." *Science* 242 (1988): 1507–1510.

Schellenberg, G. D., Bird, T. D., Wijsman, E. M., Orr, H. T., Anderson, L., Nemens, E., White, J. A., Bonnycastle, L., Weber, J. L., Alonso, M. E., Potter, H., Heston, L. L., and Martin, G. M., "Genetic Linkage Evidence For a Familial Alzheimer's Disease Locus on Chromosome 14." *Science* 258 (1992): 668–671.

Sharma, V., Smith, L., Allen, L., Magenis, R. E., and Litt, M., "Dinucleotide Repeat Polymorphism at the D14S43 Locus." *Nucleic Acids Research* 19 (1991): 1722.

Shine, I., and Wrobel, S. *Thomas Hunt Morgan.* Lexington: University of Kentucky Press, 1976.

Singer, S. *Human Genetics: An Introduction to the Principle of Heredity.* New York: W. H. Freeman and Co., 1978.

Sisodia, S. S., Koo, E. H., Beyruther, K., Unterbeck, A., and Price, D. L. "Evidence That β-Amyloid Protein in Alzheimer's Disease Is Not Derived by Normal Processing." *Science* 248 (1990): 492–495.

Sjogren, T., Sjogren, H., and Lindgren, A. G. H. "Morbus Alzheimer and Morbus Pick. A Genetic, Clinical and Patho-anatomical Study." *Acta Psychiatrica et Neurologica Scandinavia Supplement* 82 (1952).

Smith, H. O. "Nucleotides Sequence Specificity of Restriction Endonucleases." *Science* 205 (1979): 455–462.

Smith, C. A. B. "The Detection of Linkage in Human Genetics." *Journal of the Royal Statistical Society* 15 (1953): 153–192.

Solomon, E., and Bodmer, W. F. "Evolution of Sickle Variant Gene." *Lancet* 1 (1979): 923.

Sootin, H. *Gregor Mendel, Father of the Science of Genetics.* New York: Vanguard, 1959.

Southern, E. M. "Detection of Specific Sequences along DNA Fragments Separated by Gel Electrophoresis." *Journal of Molecular Biology* 98 (1975): 503–517.

St. George-Hyslop, P. H., Haines, J. L., Farrer, L. A., Polinsky, R., Van Broeckhoven, C., Goate, A., Crapper, D. R., McLachlan, H., Orr, A. C., Bruni, C., Sorbit, S., Rainero, I., Foncin, J. F., Pollen, D., Cantu, J. M., Tupler, R., Voskresenskaya, N., Mayeux, R., Growdon, J., Fried, V. A., Myers, R. H., Nee, L., Backhovens, H., Martin, J. J., Rossor, M., Owen, M. J., Mullan, M., Percy, M. E., Karlinsky, H., Rich, S., Heston, L., Montesi, M., Mortilla, M., Macmias, N., Gusella, J. F., Hardy, J. A., and other members of the FAD Collaborative Study Group. "Genetic Linkage Studies Suggest That Alzheimer's Disease Is Not a Single Homogeneous Disorder." *Nature* 234 (1990): 194–197.

St. George-Hyslop, P., Haines, J., Rogaev, E., Mortilla, M., Vaula, G., Pericak-Vance, M., Foncin, J-F., Montesi, M., Bruni, A., Sorbi, S., Rainero, I., Pinessi, L., Pollen, D., Polinsky, R., Nee, J., Kennedy, J., Macciardi, F., Rogaeva, E., Liang, Y., Alexandrova, N., Lukiw, W., Schlumpf, K., Tanzi, R., Tsuda, T., Farrer, L., Cantu, J-M., Duara, R., Amaducci, L., Bergamini, L., Gusella, J., Roses, A., and McLachlan, D. C., "Genetic Evidence For a Novel Familial Alzheimer's Disease Locus on Chromosome 14." *Nature Genetics* 2 (1992): 330–334.

St. George-Hyslop, P. H., Polinsky, A., Nee, L., Haines, J., Tanzi, R., Conneally, P., Growdon, J., Myers, R. H., Pollen, D., Drachman, D., Feldman, R., Amaducci, L., Foncin, J-F., Bruni, A., Frommet, P., and Gusella, J. F., "Search for the Familial Alzheimer's Disease Gene." *Journal of Neural Transmission* 24 (1987): 13–21.

St. George-Hyslop, P. H., Tanzi, R. E., Polinsky, R. J., Haines, J. L., Nee, L., Watkins, P. C., Myers, R. H., Feldman, R. G., Pollen, D., Drachman, D., Growdon, J., Bruni, A., Foncin, J.-F., Salmon, D., Frommelt, P., Amaducci, L., Sorbi, S., Piacentini, S., Stewart, G. D., Hobbs, W. J., Conneally, P. M., and Gusella, J. F. "The Genetic Defect Causing Familial Alzheimer's Disease Maps on Chromosome 21." *Science* 235 (1987): 885–890.

St. George-Hyslop, P. H., Tanzi, R. E., Polinsky, R. J., Neve, R. L., Pollen, D., Drachman, D., Growdon, J., Cupples, L. A., Nee, L., Myers, R. H., O'Sullivan, D., Watkins, P. C., Amos, J. A., Deutsch, C. K., Bodfish, J. W., Kinsbourne, M., Feldman, R. G., Bruni, A., Amaducci, L., Foncin, J.-F., and Gusella, J. L. "Absence of Duplication of Chromosome 21 Genes in Familial and Sporadic Alzheimer's Disease." *Science* 238 (1987): 664–666.

Strittmatter, W. J., Saunders, A. M., Schmechel, D., Pericak-Vance, M., Enghild, J., Salvesen, G. S., and Roses, A. D. "Apolipoprotein E: High Affinity Binding to β-Amyloid and Increased Frequency of Type 4 Allele in Familial Alzheimer's Disease," *Proccedings of the National Academy of Science* 90 (1993):1977–1981.

Struwe, F. "Histopathologische Untersuchungen Uber Entstehung Und Wesen Der Senilen Plaques." *Zentralblatt fur Nervenheilkunde und Psychiatrie* 122 (1929): 291–307.

Suzuki, K., Katzman, R., and Korey, S. R. "Chemical Studies on Alzheimer's Disease." *Journal of Neuropathology and Experimental Neurology* 24 (1965): 211–224.

Tanzi, R. E., Vaula, G., Romano, D. M., Mortilla, M., Huang, T. L., Tupler,

R. G., Wasco, W., Hyman, B. T., Haines, J. L., Jenkins, B. J., Kalaitsi-daki, M., Warren, A. C., McInnis, M. C., Stylianos, E., Antonarakis, S. E., Karlinsky, H., Percy, M. E., Connor, L., and Growdon, J., Crapper-McLachlan, D. R., Gosella, J. F., and St. George-Hyslop, P. H., "Assessment of Amyloid β-Protein Precursor Gene Mutations in a Large Set of Familial and Sporadic Alzheimer Disease Cases." *American Journal of Human Genetics* 51 (1992): 273–282.

Tanzi, R. E., Watkins, P. C., Stewart, G. D., Wexler, N. S., Gusella, J. F., and Haines, J. L. "A Genetic Linkage Map of Human Chromosome 21: Analysis of Recombination as a Function of Sex and Age." *American Journal of Human Genetics* 50 (1992): 551–558.

Tanzi, R. E., Haines, J. L., Watkins, P. C., Stewart, G. D., Wallace, M. R., Hallewell, R., Wong, C., Wexler, N. S., Conneally, P. M., and Gusella, J. F. "Genetic Linkage Map of Human Chromosome 21." *Genomics* 3 (1988): 129–136.

Tanzi, R. E., Bird, D., Latt, S. A., and Neve, R. "The Amyloid β-Protein Gene Is Not Duplicated in Brains from Patients with Alzheimer's Disease." *Science* 238 (1987): 666–668.

Tanzi, R. E., St. George-Hyslop, P. H., Haines, J. L., Polinsky, R. J., Nee, L., Foncin, J.-F., Neve, R. L., McClatchey, A. I., Conneally, P. M., and Gusella, J. F. "The Genetic Defect in Familial Alzheimer's Disease Is Not Tightly Linked to the Amyloid β-Protein Gene." *Nature* 329 (1987): 156–157.

Tanzi, R. E., Gusella, J. F., Watkins, P. C., Bruns, G. A. P., St. George-Hyslop, P., Van Keuren, M. L., Patterson, D., Pagan, S., Kurnit, D. M., and Neve, R. L. "Amyloid β-Protein Gene: cDNA, mRNA Distribution and Genetic Linkage Near the Alzheimer's Locus." *Science* 235 (1987): 880–884.

Terry, R. D. "Electron Microscopic Studies of Alzheimer's Disease and of Experimental Neurofibrillary Tangles." In *The Central Nervous System*, International Academy of Pathology Monograph No. 9. Baltimore: Williams and Wilkins Co., 1968.

Terry, R. D. "The Fine Structure of Neurofibrillary Tangles in Alzheimer's Disease." *Journal of Neuropathology and Experimental Neurology* 22 (1963): 629–642.

Terry, R.D., Ganatas, N. K., and Weiss. M. "Ultrastructural Studies in Alzheimer's Presenile Dementia." *American Journal of Pathology* 44 (1964): 269–297.

Tonkonogy, J., and Moak, G. "Alois Alzheimer on Presenile Dementia." *Journal of Geriatric Psychiatry in Neurology* 1 (1988): 199–206.

Trivers, R. A. "The Evolution of Reciprocal Altruism." *Quarterly Review of Biology* 46: 1971.

Van Broeckhoven, C., Backhovens, H., Cruts, M., DeWinter, G., Bruyland, M., Cras, P., and Martin, J-J., "Mapping of a Gene Predisposing to Early-Onset Alzheimer's Disease to Chromosome 14q24.3," *Nature Genetics* 2 (1992): 335–339.

Van Broeckhoven, C., Haan, J., Bakker, E., Hardy, J. A., Van Hul, W., Wehnert, A., Vegter-Van der Vlis, M., and Roos, R. A. C. "Amyloid β Protein Precursor Gene and Hereditary Cerebral Hemorrhage with Amyloidoses (Dutch)." *Science* 248 (1990): 1120–1122.

Van Duinen, S. G., Castano, E. M., Prelli, F., Gerard, T. A., Bots, A. B., Luyendijk, W., and Frangione, B. "Hereditary Cerebral Hemorrhage with Amyloidosis in Patients of Dutch Origin Is Related to Alzheimer Disease." *Proceedings of the National Academy of Science* 84 (1987): 5991–5994.

Van Broeckhoven, C., Genthe, A. M., Vandenberghe, A., Horsthemke, B., Backhovens, H., Raeymaekers, P., Van Hul, W., Wehnert, A., Gheusens, J., Cras, P., Bruyland, M., Martin, J. J., Salbaum, M., Multhaup, G., Masters, C. L., Beyreuther, K., Gurling, H. M. D, Mullan, M. J., Holland, A., Barton, A., Irving, N., Williamson, R., Richards, S. J., and Hardy, J. A. "Failure of Familial Alzheimer's Disease to Segregate with the A4-amyloid Gene in Several European Families." *Nature* 329 (1987): 153–155.

Van Nostrand, W. E., Schaimer, A. H., Farrow, J. S., Cines, D. B., and Cunningham, D. D. "Protease Nexin-2/Amyloid β-Protein Precursor in Blood Is a Platelet-Specific Protein." *Biochemical and Biophysical Research Communications* 175 (1991): 15–21.

Van Duijn, C. M., Hendrike, L., Cruts, M., Hardy, J. A., Hoffman, A., and vanBroeckhoven, C. "Amyloid Precursor Protein Gene Mutation in Early-Onset Alzheimer's Disease." *Lancet* 337 (1991): 978–979.

Vessie, P. R. "On the Transmission of Huntington's Chorea for 300 Years: The Bures Family Group." *Journal of Nervous and Mental Diseases* 76 (1932): 553–573.

Volkers, W. S., Went, L. N., and Vegter-Van Der Vlis, M. "Genetic Linkage Studies in Huntington's Chorea." *Annals of Human Genetics* 44 (1980): 75–79.

Walford, R. L., and Hodge, S.E. "HLA Distribution in Alzheimer's Disease." In *Histocompatibility Testing*, edited by P. L. Terasaki. Los Angeles: University of California Press, 1980.

Wang, Z., and Weber, J. L., "Continuous Linkage Map of Human Chromosome 14 Short Tandem Repeat Polymorphisms," *Genomics* 13 (1992): 532–536.

Watkins, P. C., Tanzi, R. E., Gibbons, K. T., Tricoli, J. V., Landes, G., Eddy, R., Shows, T. B., and Gusella, J. F. "Isolation of Polymorphic DNA Segments from Human Chromosome 21." *Nucleic Acids Research* 13 (1985): 6075–6088.

Watson, J. D., Tooze, J., and Kurtz, D. T. *Recombinant DNA: A Short Course*. New York: Scientific American Books, 1983.

Watson, J. D. "The Human Genome Project: Past, Present and Future." *Science* 248 (1990): 44–49.

Watson, J. D., and Crick, F. H. C. "Molecular Structure of Nucleic Acids: A Structure for Deoxyribose Nucleic Acid." *Nature* 171 (1953): 737–738.

Weber, J. L., and May, P. E. "Abundant Class of Human DNA Polymorphisms Which Can Be Typed Using the Polymerase Chain Reaction." *American Journal of Human Genetics* 44 (1989): 388–396.

Weitkamp, L. R., Nee, L., Keats, B., Polinsky, R. J., and Guttormsen, S., "Alzheimer Disease: Evidence for Susceptibility Loci on Chromosomes 6 and 14," *American Journal of Human Genetics* 35 (1983): 443–453.

Wexler, N. "Life in the Lab." *Los Angeles Times Magazine,* February 10, 1991, pp. 12–13, 31.

Whalley, L. J., Urbaniak, S. J., Daarg, C., Peutherer, J. F. and Christie, J. E. "Histocompatibility Antigens and Antibodies to Viral and Other Antigens in Alzheimer's Presenile Dementia." *Acta Psychiatrica Scandinavia* 61 (1980): 1–7.

Wheelan, L., and Race, R. R. "Familial Alzheimer's Disease." *Annals of Human Genetics* 23 (1959): 300–310.

Wiggins, S., Whyte, P., Huggins, M., Adam, S., Theilmann, J., Bloch, M., Sheps, S. B., Schechter, M. T., and Hayden, M. R., "The Psychological Consequences of Predictive Testing For Huntington's Disease," *The New England Journal of Medicine* 327 (1992): 1401–1405.

Wilson, E. O. *Sociobiology: The New Synthesis.* Cambridge, Mass: The Belknap Press of Harvard University, 1975.

Wolstenholme, G. E. W., and O'Connor, M. *Ciba Symposium on Alzheimer's Disease and Related Conditions.* London: Churchill, 1979.

Wong, C. W., Quaranta, V., and Glenner, C. G. "Neuritic Plaques in Cerebrovascular Amyloid in Alzheimer's Disease Are Antigenically Related." *Proceedings of the National Academy of Science* 82 (1985): 8729–8732.

Wyman, A. R., and White, R. L. "Construction of a Genetic Linkage Map in Man Using Restriction Fragment Length Polymorphisms." *American Journal of Human Genetics* 32 (1980): 314–331.

Wyman, A. R., and White, R. "A Highly Polymorphic Locus in Human DNA." *Proceedings of the National Academy of Science* 77 (1980): 6754–6758.

Yankner, B. A., Duffy, L. K., and Kirschner, D. A. "Neurotrophic and Neurotoxic Effects of Amyloid β-Protein: Reversal by Tachykinin Neuropeptides." *Science* 250 (1990): 279–282.

Young, A. W. "Franz Nissl and Alois Alzheimer." *Archives of Neurology and Psychiatry* 33 (1935): 847–852.

Zigas, Vincent. *Laughing Death. The Untold Story of Kuru.* Clifton, N.J.: Humana, 1990. (See also review of book by Len Goodwin in *Nature* 346 (1990): 28.)

Index